OTHER TITLES OF INTEREST FROM ST. LUCIE PRESS

The Motivating Team Leader

The New Leader: Bringing Creativity and Innovation to the Workplace

Best Team Skills: 50 Key Skills for Unlimited Team Achievement

The Baldrige Award for Education: How to Measure and Document Quality Improvement

Teams in Education: Creating an Integrated Approach

Organization Teams: Building Continuous Quality Improvement

Reengineering the Training Function: How to Align Training with the New Corporate Agenda

Team Building: A Structured Learning Approach

Real Dream Teams: Seven Practices Used by World-Class Team Leaders to Achieve Extraordinary Results

Leadership by Encouragement

The Skills of Encouragement: Bringing Out the Best in Yourself and Others

What Is, Is: Encouraging Yourself to Accept What You Can't Change and to Change What You Can

Teams in Government: A Handbook for Team-Based Organizations

Mastering the Diversity Challenge: Easy On-the-Job Applications for Measurable Results

For more information about these titles call, fax or write:

St. Lucie Press
2000 Corporate Blvd., N.W.
Boca Raton, FL 33431-9868

TEL (561) 994-0555 • (800) 272-7737
FAX (800) 374-3401
E-MAIL information@slpress.com
WEB SITE http://www.slpress.com

SₜL

TQM
Facilitator's Guide

TQM

Facilitator's Guide

Jerome S. Arcaro

S^t_L

St. Lucie Press
Boca Raton, Florida

Phone: (561) 994-0555
E-mail: information@slpress.com
Web site: http://www.slpress.com

S$_L^t$

Published by
St. Lucie Press
2000 Corporate Blvd., N.W.
Boca Raton, FL 33431-9868

Contents

Preface

The Galileo Quality Institute was founded by professionals with experience in quality management, education, business, and government. Our goal is to help improve business practices, cost effectiveness, and customer satisfaction. Our mission is to help every organization improve the quality of products and services. We firmly believe that today's business, education, and government professionals possess the expertise necessary to improve business processes and operational results. Working with our partners, we have developed a creative framework known as our Excellence In Organizational Management System™ to bring Continuous Improvement processes to the fields of business, education, and government. The Institute's organizational capacity-building programs are currently being implemented in the United States and in Europe.

Guided by our attention to details, people, and training, we promote high performance standards that lead to positive reinforcement and continuous improvement. Our programs embody the fine art of perfecting business fundamentals. Using a range of internal and external resources, the Galileo Quality Institute is uniquely prepared to implement TQM-based programs in business, education, and government. Our Excellence In Organizational Management System™ is internationally recognized for its focus on improving every aspect of the organization, from meeting customer requests to improving the physical condition of the work environment.

The methodology discussed in this handbook was developed over a number of years. Facilitation services have been and are currently being provided to organizations in the United States and Europe. As you will learn in this handbook, facilitation styles differ according to the culture of the organization. For example, in England team structure is based on organizational classification. Team leaders are usually selected from the management team. Teams also use very formal processes for managing team meetings. In this environment, the facilitator provides assistance to the team only when he/she is asked to do so. The facilitator does not volunteer any assistance to the team.

Currently, Galileo is providing facilitation services to public and private organizations in the United States. For example, it is providing facilitation services to the Southern Division, Naval Facilities Engineering Command and the U.S. Air Force in Ohio. These facilitation services support the services' Environmental Resolution Teams. The teams are comprised of representatives from the service, the base, the state Environmental Protection Agency, the U.S. Environmental Protection Agency, and the contractors hired to perform the clean-up activities.

These teams work in a very informal environment. The facilitators are encouraged to volunteer assistance when the teams become entangled in controversy or when the team cannot reach a consensus. A major role of the facilitator is to provide the team with team-building, problem-solving, and conflict-resolution training. In this environment, the facilitator is a proactive resource for the team.

Jerome Arcaro
Galileo Quality Institute

If you have any questions regarding the material in this book, please contact the Galileo Quality Institute at Unit 51 - Pine Street Extension, Nashua, NH 03060. Our telephone number is (800) 242-1464, or you may fax your inquiries to us at (603) 883-2330.

Author

TQM Facilitator's Guide was developed by a professional who has extensive experience in implementing Total Quality Management in business, education, and government. Mr. Jerome S. Arcaro, President, Galileo Quality Institute, has developed the Institute's internationally acclaimed "Quality Assessment System for Businesses, Schools, and Government Agencies" (QAA). QAA is a tool being used by business, education, and government professionals to measure organizational effectiveness, customer satisfaction, and supplier performance. Mr. Arcaro has implemented cultural change initiatives in organizations in the United States and in England.

Mr. Arcaro is the author of *Quality in Education: An Implementation Handbook, Teams in Education: Creating an Integrated Approach*, and *The Baldrige Award for Education*. This book was developed by Mr. Arcaro to help you improve your team facilitation skills. Mr. Arcaro is a guest speaker at business, education, and government forums. He has conducted more than 300 public seminars in the United States and in England.

Chapter 1

Identifying Your Motivation to Become a Facilitator

Name: _____ Position: _____

1. Why did you volunteer to become a facilitator, or why do you want to develop facilitation skills?

_____

_____

_____

_____

_____

_____

2. If you have been (or are) a facilitator, please describe what you did (or do) and with what types of groups.

_____

_____

_____

_____

_____

_____

_____

3. If you have had previous facilitator training, please describe.

_____

_____

_____

_____

_____

_____

_____

4. If not described in question #1, what specific skills, knowledge, etc. do you want to learn?

_____

_____

_____

_____

_____

_____

5. What concerns you the most about becoming a facilitator or learning facilitation skills?

_____

_____

_____

_____

_____

_____

_____

6. **What questions do you have?**

_____

_____

_____

_____

_____

_____

_____

_____

7. **What are your qualifications for being a facilitator?**

_____

_____

_____

_____

_____

_____

_____

8. **Have you had any group dynamics training?**

_____

_____

_____

_____

_____

_____

_____

9. How much time do you expect to devote to your facilitator responsibilities?

_____

_____

_____

_____

_____

_____

_____

10. What are your comments and concerns?

_____

_____

_____

_____

_____

_____

_____

11. Do you have any suggestions relative to the type of training you need?

_____

_____

_____

_____

_____

_____

_____

12. Describe how you will benefit from this experience.

_____

_____

_____

_____

_____

_____

_____

_____

13. Do you know anyone else who might like to be a facilitator?

_____

_____

_____

_____

_____

_____

_____

_____

14. Will you devote after-work hours to being a facilitator?

_____

_____

_____

_____

_____

_____

_____

_____

15. Are you willing to train other people to become facilitators?

_____

_____

_____

_____

_____

_____

_____

_____

Are You Properly Motivated to Become a Facilitator?

Chapter 2

Pre-Course Self-Assessment

Name: _____   Position: _____

The following is a list of skills and knowledge that will be covered in this workbook. If you feel reasonably comfortable with your level of skill or knowledge, place a check next to the item. If not, leave it blank. You should complete the self-assessment before using this workbook. It will help you to identify the topics requiring additional focus.

1. ___ Differences between facilitating, leading, and training

2. ___ Roles and responsibilities of the facilitator

3. ___ Roles people play in groups

4. ___ Adult learning principles

5. ___ Formation of groups

6. ___ Group member responsibilities

7. ___ Group problems and dynamics

8. ___ TQM principles and practices

9. ___ Decision making

10. ___ Union or association roles

11. ___ Conflict management

12. ___ Effective meetings

13. ___ Management presentations

14. ___ Problem-solving models

15. ___ Customer identification and relationships

16. ___ Brainstorming

17. ___ Cause-Effect Diagrams (Fishbone)

18. ___ Checksheets

19. ___ Cost-Benefit Analysis

20. ___ Force-Field Analysis

21. ___ List reduction techniques

22. ___ Matrix Diagrams

23. ___ Pareto Charts

24. ___ Surveys and Interviews

25. ___ PDCA

26. ___ PERT Charts

27. ___ GANTT Charts

28. ___ Listening

29. ___ Behavior observation and analysis

30. ___ TQM resources

31. ___ Team building

32. ___ Motivation

33. ___ Time management

34. ___ Flowcharts

35. ___ Histograms

36. ___ Mind Maps

37. ___ Run Charts

38. ___ Trust building

39. ___ Negotiation and mediation

40. ___ Strategic planning

41. ___ Organizational development

42. ___ Organizational diagnosis

43. ___ Process observing

44. ___ Affinity Diagrams

45. ___ Interrelationship Diagraphs

46. ___ Tree Diagrams

47. ___ Prioritization Matrices

48. ___ Process Decision Program Charts

49. ___ Activity Network Diagrams

Chapter 3

Facilitation

In today's competitive society, organizations must constantly be on the alert for new ideas and practices that will give them the edge. Over the past twenty years, organizational capacity-building engineers (my definition of facilitators) have helped teams free their creative talents to develop new and innovative ways to work. This has provided organizations with numerous benefits that could not have been achieved without the guidance and support of a trained facilitator.

This workbook takes you on a mind-opening journey into the science of facilitation. Structured as a self-directed learning workbook, the reader develops and builds on existing team skills. This practical handbook is a must for facilitators who want a down-to-earth approach to creating, developing, and growing successful teams.

You will discover that this workbook will help you guide teams from an environment of inaction, fear, and miscommunication to an environment that is fertile with discussion, understanding, and cooperation. You will learn new skills that will help you guide your teams to the development of a shared dedication to quality.

What Is Facilitation?

Facilitating (to make easy or less difficult) is probably the most misunderstood and misused term in the entire Total Quality Management (TQM) process, team-building initiatives, cultural and organizational change initiatives, and organizational capacity-building initiatives. Because of this, the role of the facilitator is confusing and even contradictory. Who is a facilitator? What is the role of the facilitator? What groups are facilitated? These are only some of the questions that we are most frequently asked.

What Type of Person Becomes a Facilitator?

A facilitator is usually a volunteer from within the organization who receives training to help groups work more effectively. This function is usually performed in addition to the facilitator's regular job. In most situations, released time is provided to do facilitating and, in addition, compensation is sometimes provided in recognition for the extra time commitment and training involved.

An external facilitator is someone outside the organization. An external facilitator is usually a consultant who is assisting in the organization's quality transformation or capacity-building initiative. One of the major goals for every external facilitator is to eliminate his/her job. External facilitators must provide training to staff that enables the organization to develop its own internal facilitators.

What Is the Role of the Facilitator?

The literature has the facilitator filling a number of different roles. A facilitator must be trained as a:

- change agent
- organizational development consultant
- mediator and negotiator
- trainer and educator
- team builder
- process observer
- TQM expert
- problem solver
- statistician
- listener
- communicator

What Groups Are Facilitated?

There are many different types of teams that require the services of a facilitator. Quality teams may be long-standing, temporary, or on-going. The following are some of the names of the quality teams that we have worked with:

- task teams
- planning teams
- design teams
- functional teams
- department teams
- administrative teams

- **cross-functional teams**
- **self-directed teams—if requested**
- **councils**
- **committees**
- **boards**

To require or expect all facilitators to be all things to all types of groups is unrealistic and ineffective, and it can even be counter-productive. The facilitator is not a member of the group, does not participate in group activities, and has no authority. The authority and responsibility rest with the team leader. In addition, the facilitator should not have any close working relationships with the team or its members. One of the fundamental requirements for facilitating is objectivity and neutrality.

However, the reality is that in a small organization, it is not always possible to have completely neutral facilitators. Most likely, there are working relationships with team members or departments within the organization. This makes the job of the facilitator more difficult.

What Is the Difference Between a Team Leader and a Facilitator?

The role difference between the team leader and the facilitator is that the team leader is responsible for getting the team's task accomplished and the facilitator is responsible for ensuring that the process involved is honest and effective.

The facilitator must develop an effective working relationship with the team leader to ensure that the team leader does not feel threatened by the presence of the facilitator. Facilitators overcome this problem because they are trained in the process of facilitation and team management. Generally, team leaders do not receive any training. In many instances, team leaders do not know what is expected of them. This is a recipe for conflict!

There are only two acceptable solutions to this problem: either the team leader receives training in the role and use of a facilitator, or the facilitator provides the team leader with required training. The team cannot be successful unless the team leader is prepared to perform the duties expected of him/her. The team leader must receive training in team management processes.

Some organizations adopt another possible solution, but I do not recommend it. The team leader can act as both the facilitator and team leader. If a team requires the services of a facilitator, then the roles should be kept separate. Based on my experience, cross-functional teams that have the team leader serve in both roles fail because the team leader has a vested interest in solving the problem. It is extremely difficult for the other members of the team to view the team leader as being impartial and objective.

The leader is usually not known until the team is formed. There are exceptions where organizational practices, policies, or positions dictate who the team leader must be. Quite often it is an administrator or manager. Therefore, it would be a good strategy to have administrators and managers trained in facilitation skills because they are valuable and even necessary skills to have.

What Are the Characteristics of an Effective Facilitator?

Although anyone can strive to become a facilitator, certain basic characteristics are helpful. A facilitator should:

- like working with people
- be flexible (not strong-willed or opinionated)
- enjoy instructing/teaching
- communicate effectively
- be interested in learning
- handle criticism constructively

There are organization-specific qualifications that a facilitator must have. In some organizations, the facilitator must:

- be a subject-matter expert
- have achieved a certain level within the organization
- have knowledge of specific issues impacting the organization
- be a certified facilitator

Exercise

In the space below list the facilitator characteristics that are required for your organization.

_____

_____

_____

_____

_____

_____

_____

_____

What Knowledge and Skills Are Required to Be a Successful Facilitator?

At Galileo, we classify facilitators according to their skill level. On the basis our past experience, we created three facilitator classifications: Facilitator I, Facilitator II, and Facilitator III. A facilitator is required to have basic knowledge and skills in order to be accepted, trusted, and effective.

Facilitator I

A *Facilitator I* possesses the basic skills needed to keep a performing team focused on its task. In this instance, the facilitator plays the role of an evaluator. The facilitator provides the team with feedback relative to its performance, efficiency, and effectiveness. A Facilitator I is never assigned to a dysfunctional team. Usually, a Facilitator I is assigned to high-performing or function teams.

A *Facilitator I* must have a working knowledge of:

- a facilitator's role and responsibilities
- different learning styles
- a group leader's role and responsibilities
- group/team formation
- group/team roles and responsibilities
- group problems
- union or association roles
- Total Quality Management/Total Quality Systems (TQM/TQS) basic principles and practices
- decision making
- TQM resources

Additionally, a Facilitator I must also possess and be able to effectively demonstrate skills in the following areas:

◊ **Team Meeting Management Skills**

- organizing team meetings
- creating cross-functional teams
- developing team problem statements
- defining team issues
- creating a customer focus
- getting meetings started
- conducting effective team meetings
- resolving conflict between team members
- developing effective management presentations

◊ **Quality Management Tools and Techniques**

- Brainstorming
- Cause-Effect Diagrams
- Checksheets
- Cost-Benefit Analysis
- Force-Field Analysis
- List Reduction
- Matrix Diagrams
- Pareto Charts
- Surveys and Interviews

◊ **Planning and Monitoring**

- quality planning
- PDCA
- PERT Charts
- GANTT Charts

◊ **Evaluation**

- quality standards and indicators

◊ **People Process**

- listening

- observing

- providing positive reinforcement

- handling disruptive behaviors

Facilitator II

A *Facilitator II* has extensive experience and training in understanding the psychological aspects of team development. Usually, a Facilitator II is assigned to a non-performing team. Non-performing teams experience difficulty in developing a customer focus, the team members cannot define the issues impacting the team, and the team cannot reach a consensus on basic issues. The team does not accomplish its task.

In addition to the skills and knowledge previously discussed, a Facilitator II must possess the following:

◊ **Knowledge**

- adult learning theory

- motivation

- group dynamics

- sources of power

- team development

- TQM/TQS models

- time management

◊ **Skills**

- quality process models
- core focus
- management information systems
- systems theory
- dynamics of team building
- establishing and building team trust

◊ **Quality Tools and Techniques**

- Environmental Scanning
- Flowcharts
- Histograms
- Mind Mapping
- Run Charts
- Seven Management and Planning Tools

◊ **People Processes**

- handling resistance
- handling disruptive people
- negotiating and mediating

Facilitator III

A *Facilitator III* has received extensive training in organizational development, team building, conflict resolution, group dynamics, problem resolution, and communication. A Facilitator III has extensive experience in facilitating performing and non-performing teams. A Facilitator III is a very mature individual who is well-respected throughout the organization.

Usually, a Facilitator III is assigned to work the organization's most critical teams. These teams are highly visible throughout the organization. Critical teams focus on the core issues that drive the business. In most cases, membership on these teams is reserved for the most senior executives.

In addition to the knowledge and skills previously discussed, a Facilitator III must possess the following:

◊ **Knowledge**

- organizational development
- TQM/TQS principles and practices
- personality type testing
- counseling

◊ **Skills**

- organizational diagnosis
- strategic quality planning
- communication
- research
- public presentations
- selling

To become a registered facilitator, complete and return the Facilitator Skills Assessment Form to the Galileo Quality Institute, Unit 51-Pine Street Extension, Nashua, NH 03060 or fax to (603) 883-2330.

Description of Facilitator Competencies

Analytical Thinking

Discriminates between important and unimportant details, recognizes inconsistencies between facts, and draws inferences from information.

Forecasting

Accurately anticipates changes in workloads, resources, and personnel needs as a result of changes in the work situation, technology, or external developments.

Goal Orientation

Ensures that the results to be achieved by the team or individual are clearly defined and understood at all times.

Knowledge of User Support Areas

Has a basic understanding of the various user areas being supported, their needs, and their technical requirements.

Multiple Focus

Effectively manages a large number of people with different and often conflicting objectives, projects, groups, or activities at one time.

Organizational Knowledge

Has a thorough understanding of organizational policies, procedures, and key personnel that enable a person to effectively carry out his/her responsibilities.

Priority Setting

Identifies and separates those tasks that are most important from those that are less important; maintains a clear sense of priorities and a vision of the larger picture.

Risk Taking

Takes risks when the consequences are difficult to predict but the payoffs are likely to be great, even when proposals may be rejected by superiors or when one's image may suffer if wrong.

Facilitator Skills Assessment

Part I

According to the following scale, please rank your knowledge of and expertise in the following areas:

ITEM	LOW	MEDIUM	HIGH
Priority setting	____	____	____
Organization and planning	____	____	____
Multiple focus	____	____	____
Decisiveness	____	____	____
Communication skills—oral	____	____	____
Selection	____	____	____
Delegation	____	____	____
Organizational knowledge	____	____	____
Analytical thinking	____	____	____
Motivation	____	____	____
Fact finding	____	____	____
Goal orientation	____	____	____
Strategic thinking	____	____	____
Team building	____	____	____
Forecasting	____	____	____
Management efficiency	____	____	____
Systems perspective	____	____	____
Negotiation	____	____	____
Developing others	____	____	____
Risk taking	____	____	____
Leadership	____	____	____
Time management	____	____	____
Business management	____	____	____
Conflict resolution	____	____	____
Flexibility	____	____	____

Facilitator Skills Assessment

Part II

According to the following scale, please rank your knowledge of and expertise in the following areas:

ITEM	LOW	MEDIUM	HIGH
Accessibility	____	____	____
Analytical thinking	____	____	____
Assertiveness/autonomy	____	____	____
Attention to detail	____	____	____
Coaching	____	____	____
Communications—written	____	____	____
Conducting meetings	____	____	____
Control systems	____	____	____
Coordination	____	____	____
Decisiveness	____	____	____
Energy level	____	____	____
Evaluation	____	____	____
Feedback	____	____	____
Flexibility	____	____	____
Innovation	____	____	____
Meeting management	____	____	____
Motivating others	____	____	____
Networking	____	____	____
Organizing and planning	____	____	____
Participation	____	____	____
Persistence	____	____	____
Personality types	____	____	____
Quality principles	____	____	____

Facilitator Skills Assessment

Part III

According to the following scale, please rank your knowledge of and expertise in the following areas:

ITEM	LOW	MEDIUM	HIGH
Team preparation	_____	_____	_____
Meeting management	_____	_____	_____
Developing problem statements	_____	_____	_____
Effective use of resources	_____	_____	_____
Accuracy in forecasting	_____	_____	_____
Timeliness	_____	_____	_____
Follow-through	_____	_____	_____
Training	_____	_____	_____
Being a neutral observer	_____	_____	_____
Defining the problem	_____	_____	_____
Determining competencies	_____	_____	_____
Creating a customer focus	_____	_____	_____
Developing a team vision	_____	_____	_____
Task focused	_____	_____	_____
Critical thinking	_____	_____	_____
Building expertise	_____	_____	_____
Survival skills	_____	_____	_____
Defining learning objectives	_____	_____	_____
Developing training programs	_____	_____	_____
Adult learning needs	_____	_____	_____
Organization development styles	_____	_____	_____
Keeping teams focused	_____	_____	_____
Conflict avoidance	_____	_____	_____
Valuing cultural differences	_____	_____	_____
Assessment	_____	_____	_____

Facilitator Classification Form

Please complete the following. It will help you identify your facilitator classification. You can also use this information to enhance your facilitation skills. You do not have to share this information with anyone.

ITEM	AREA OF STRENGTH	AREA FOR IMPROVEMENT
Facilitator's role and responsibilities	_____	_____
Different learning styles	_____	_____
Group leader role and responsibilities	_____	_____
Group/team formation	_____	_____
Group/team roles and responsibilities	_____	_____
Group problems	_____	_____
Union or association roles	_____	_____
TQM/TQS basic principles and practices	_____	_____
Decision making	_____	_____
TQM resources	_____	_____
Organizing team meetings	_____	_____
Creating cross-functional teams	_____	_____
Developing team problem statements	_____	_____
Defining team issues	_____	_____
Creating a customer focus	_____	_____
Getting meetings started	_____	_____
Conducting effective team meetings	_____	_____
Resolving conflict	_____	_____
Developing management presentations	_____	_____
Adult learning theory	_____	_____
Motivation	_____	_____
Group dynamics	_____	_____
Sources of power	_____	_____
Team development	_____	_____
TQM/TQS models	_____	_____
Quality process models	_____	_____

Chapter 4

Facilitator Roles and Responsibilities

Basically, the role of the facilitator is to make it easier for a group to do problem solving, make decisions, and address issues and to manage change rather than having change manage the group.

In order to accomplish this very difficult and delicate task, the facilitator needs to be:

- neutral
- objective
- positive
- instructive
- encouraging
- resourceful
- inventive and imaginative
- protecting
- focused
- knowledgeable
- skillful
- trusted
- informative
- communicative
- honest

Another responsibility of the facilitator is to inform a group about its role and responsibility. Sometimes it is assumed that a group knows what a facilitator does, but this is simply not the case.

The facilitator should be very open with the group about the fact that it is a difficult and delicate job, that the role is to help the group, and that feedback is appreciated.

Do You Have the Skills?

Please check the appropriate box for each item. This will help you to focus your efforts in the areas that need improvement.

Facilitator Skills Self-Assessment

ITEM	AREA OF STRENGTH	AREA FOR IMPROVEMENT
Neutral observer	_____	_____
Objective	_____	_____
Positive attitude	_____	_____
Instructive	_____	_____
Encouraging	_____	_____
Resourceful	_____	_____
Inventive and imaginative	_____	_____
Protecting	_____	_____
Focused	_____	_____
Knowledgeable	_____	_____
Skillful	_____	_____
Trusted	_____	_____
Informative	_____	_____
Communicative	_____	_____
Honest	_____	_____
Adult learning needs	_____	_____

The adult learner, unlike a young learner, has a fundamental need to be far more participative in a learning and work environment because mental growth and development is a natural and necessary function of adult living. Adults have a need to:

- feel valued
- feel competent
- feel secure
- do interesting work
- receive fair treatment
- participate in decisions affecting them
- validate information based on their personal beliefs and experiences
- learn what will be immediately useful
- be praised and rewarded
- be understood
- be appreciated
- feel part of a group
- have an identity
- use their talents and skills
- draw upon their experience and interests

Translating the basic adult needs to the workplace, adult learners want more:

- responsibility
- control over their environment
- decision-making power
- influence
- information
- skills and training

What is particularly and significantly interesting is that this is exactly the type of workers organizations need to have today if they are to be more effective and if they are to maximize their human resources. The reality is that quality must be an obsessive goal because customers have higher expectations and requirements. If organizations are to be more effective, they must become more responsive more quickly. For this to occur, the following changes need to take place:

- encouraging working smarter, not harder
- working in teams and in collaboration and cooperation
- creating an environment for innovation and change
- empowering workers to respond quickly to customer needs
- developing flexible and multi-skilled workers
- increasing involvement and responsibility for continuous improvement
- establishing clear goals and expectations
- maintaining shared and practiced beliefs and values

In contrast, what adults do not need is to feel:

- controlled
- left out of things
- blamed and/or criticized
- incompetent
- insecure
- misunderstood or ignored
- taken for granted
- overloaded
- bored

Teaming is so valuable for adults because it makes them feel a part of what is going on in the workplace by participating and contributing their skills and abilities.

Team Leader Roles and Responsibilities

The leader of the group, whether elected or appointed, has the clear role of leading the group to accomplish the task assigned so that the group feels good about the final result. It is possible to accomplish the task but not have a group feel good about the process—this is the reason a facilitator is necessary.

Thus, leaders have the responsibility to do the following:

- act consistently
- eliminate fear
- develop trust
- give constructive feedback
- give praise and recognition
- provide whatever information is necessary
- model tolerance and flexibility
- create a respectful environment
- encourage creativity and risk taking
- keep the group focused on the task
- protect members from personal attack
- encourage participation and discussion
- be clear about how decisions will be made
- develop shared guidelines for the group process
- encourage and support facilitator assistance
- follow guidelines for effective meetings
- evaluate group effectiveness

Do You Have the Skills?

Please check the appropriate box for each item. This will help you to focus your efforts in the areas that need improvement.

Team Leader Roles and Responsibilities Skills Self-Assessment

ITEM	AREA OF STRENGTH	AREA FOR IMPROVEMENT
Act consistently	_____	_____
Eliminate fear	_____	_____
Develop trust	_____	_____
Give constructive feedback	_____	_____
Give praise and recognition	_____	_____
Provide whatever information is necessary	_____	_____
Model tolerance and flexibility	_____	_____
Create a respectful environment	_____	_____
Encourage creativity and risk taking	_____	_____
Keep the group focused on the task	_____	_____
Protect members from personal attack	_____	_____
Encourage participation and discussion	_____	_____
Be clear about how decisions will be made	_____	_____
Develop shared guidelines for group process	_____	_____
Encourage and support facilitator assistance	_____	_____
Follow guidelines for effective meetings	_____	_____
Evaluate group effectiveness	_____	_____

Group/Team Formation

Studies have indicated that group participation and problem solving are far more effective than an individual effort. The reason is really quite simple: work tasks can only be accomplished with and through other people. If this is not done effectively, the following messages can be heard very loudly, even if not uttered verbally.

"We always seem to be going in different directions."

"Why do we waste time in meetings? Nothing is ever really accomplished."

"No one knows what is going on around here."

"Plans just don't get executed; something always falls through the cracks."

"There seems to be so much duplication of effort."

"Everyone is busy, but nothing seems to get done."

"Some people don't seem very enthusiastic."

"Why is there so much grumbling around here?"

"Communication is really a problem around this place."

"Why is someone always inspecting what we do; don't they trust us?"

"Why should I bust my back; it's never appreciated."

"Be careful what you say around here."

Add Some of Your Own

_____

_____

_____

_____

_____

_____

_____

_____

_____

_____

_____

_____

_____

_____

However, simply having a group work together does not ensure effectiveness unless the group is properly structured and instructed. In the Appendix, there is a form for outlining the guidelines required whenever a group, team, or committee is formed.

Group Member Responsibilities

A group leader cannot be effective, nor can the group's task be achieved successfully, unless the members of the group also fulfill their responsibilities, which include the following:

- supporting and assisting the group leader
- participating by expressing opinions and feelings
- supporting one another
- maintaining confidentiality (if this is one of the guidelines)
- showing loyalty to the organization
- criticizing constructively
- suggesting options or alternatives
- adhering to the guidelines
- not allowing or participating in personal attacks
- not monopolizing the discussion
- coming prepared
- keeping the discussion focused on the task and agenda
- providing data to support opinions
- volunteering for extra assignments
- listening to what other members are saying
- trying to understand diverse opinions
- being willing to "give and take"
- asking questions
- seeking clarification
- not using inflammatory words
- treating everyone with respect
- avoiding hidden agendas

How Do Team Members in Your Organization Perform?

For the teams that you have been a member of or for the teams that you have facilitated, please indicate your assessment of the team members' performance by checking the appropriate box.

Team Member Roles and Responsibility Self-Assessment

ITEM	NEVER PRACTICED	USUALLY PRACTICED	ALWAYS PRACTICED
Support and assist the group leader	_____	_____	_____
Participate by expressing opinions/feelings	_____	_____	_____
Support others	_____	_____	_____
Maintain confidentiality	_____	_____	_____
Show loyalty to the organization	_____	_____	_____
Criticize constructively	_____	_____	_____
Suggest options or alternatives	_____	_____	_____
Adhere to the guidelines	_____	_____	_____
Do not allow personal attacks	_____	_____	_____
Do not monopolize the discussion	_____	_____	_____
Come prepared	_____	_____	_____
Keep the discussion focused on the task	_____	_____	_____
Provide data to support opinions	_____	_____	_____
Volunteer for extra assignments	_____	_____	_____
Listen to what other members are saying	_____	_____	_____
Try to understand diverse opinions	_____	_____	_____
Willing to "give and take"	_____	_____	_____
Ask questions	_____	_____	_____
Seek clarification	_____	_____	_____
Do not use inflammatory words	_____	_____	_____
Treat everyone with respect	_____	_____	_____
Do not bring hidden agendas	_____	_____	_____

Exercise

In the space below, list the roles and responsibilities for the facilitators in your organization.

Roles People Play Here

_____

_____

_____

_____

_____

_____

_____

_____

_____

_____

_____

_____

_____

Team Issues

In the space below, list the team issues that might be encountered in your organization.

_____

_____

_____

_____

_____

_____

_____

_____

_____

_____

_____

_____

_____

_____

_____

_____

_____

Symptoms of Team Issues

In the space below, list the symptoms of the team issues that might be encountered in your organization.

_____

_____

_____

_____

_____

_____

_____

_____

_____

_____

_____

_____

_____

_____

_____

Union/Association Roles

In the early 1980s, a quality movement took place in business and industry, with Quality Circles and Quality of Work Life Programs being implemented widely. Both efforts addressed the need to consider the role of the unions in the process.

The TQM effort is not addressing this issue. One reason may be that union membership has declined in the business and industrial sectors; however, it has increased substantially in the public sector. Therefore, the role of the union must be considered very carefully before proceeding with any quality effort, particularly with the teaming aspect. Any attempt to circumvent the union role or to even give the appearance of interfering with union responsibilities can lead to unnecessary conflict.

Respect must be given to the role of the union in protecting its membership and negotiating responsibilities, which include working conditions.

The best way to approach this issue is to have a representative from each union be an integral part of the quality process. In this way, if any issue comes up that a union may find violates its role and/or responsibility, that issue can be addressed quickly and within the TQM process.

Are You Ready to Move Forward?

Chapter 5

Basic Principles of Total Quality Management

Quality and continuous improvement are the most important issues in business today. We all recognize that there are problems with today's business systems. The quality of many our nation's products and services is getting better. However, in many instances, the quality of these products and services is consistently below the quality standards established by foreign competition.

In many instances, workers are not prepared to accept the concept of team activities. Workers are not trained in the use of problem-solving tools and techniques. Customer attitudes have changed. Customers are demanding products and services that consistently meet their requirements. Failure to respond to market conditions has resulted in American businesses losing market share to foreign competition and employees losing jobs to foreign workers.

TQM, Continuous Improvement, and Quality Improvement Teams are some of the vehicles that business professionals are using to cope with the "forces of change" that are buffeting our nation's business systems. The knowledge needed to improve our business environment already exists within the business community. The major difficulty business professionals face today is their inability to deal with the "system failures" that are preventing them from developing or implementing new work processes which will improve the quality of the products and services. Everyone recognizes that quality improvements result in increased profitability and improved customer satisfaction.

Business must undergo a paradigm shift. Old norms and beliefs must be challenged. Workers must now compete in a global economy. Business must learn to work with fewer resources. Many of today's business professionals lack the knowledge or expertise necessary to prepare workers to adapt to changing labor market requirements. Tradition prevents many business processes from being changed to meet customer needs. Society is demanding that the quality of products and services improve, but society fails to support businesses' efforts to improve. Many of our nation's business professionals are fearful of change and don't know how to cope with the new requirements expected of them.

The modern world is one in which the only constant is change. Change is a precarious matter, but TQM can help businesses cope with change in a positive and constructive manner. A quick fix will not solve today's business problems. It will take dedication and focus and a constancy of purpose on behalf of all of the stakeholders to achieve a quality work environment.

TQM is a tool that can help business professionals cope with today's changing environment. TQM helps to alleviate fear and increase trust in businesses. TQM provides focus for business. It establishes a flexible infrastructure that can quickly respond to the customer's changing demands. It helps business to cope with budget and time constraints. *TQM makes it easier to manage change.*

The transformation to a Total Quality Business begins with the adoption of a shared dedication to quality by the executives, workers, customers, and suppliers. The process begins with the development of a Quality Vision and Mission for the organization and for functional units within the organization. The Quality Vision focuses on meeting the needs of the customers, on providing for total business involvement in the program, on developing systems to measure the added value of the process, on support systems that managers and workers need to manage change, and on continuous improvement, always striving to make products or services better.

The basic principles of TQM are:

- shared beliefs and values
- shared vision and mission
- teamwork and collaboration
- constancy of purpose
- consistent message and behavior
- systemic improvements
- prevention rather than inspection
- data-based decisions
- on-going education and training
- pride of workmanship
- joy in learning
- information sharing
- problem solving
- innovation
- partnership development
- customer focus
- quality priority
- fixing the system rather than fixing blame
- process emphasis
- cost-benefit analysis
- value-added view
- continuous improvement
- elimination of waste
- reduction of waste
- valuing people

In the following worksheet, indicate the degree to which quality principles are practiced in your organization.

Total Quality Principles Self-Assessment

ITEM	NEVER PRACTICED	USUALLY PRACTICED	ALWAYS PRACTICED
Shared beliefs and values	___	___	___
Shared vision and mission	___	___	___
Teamwork and collaboration	___	___	___
Constancy of purpose	___	___	___
Consistent message and behavior	___	___	___
Systemic improvements	___	___	___
Prevention rather than inspection	___	___	___
Data-based decisions	___	___	___
On-going education and training	___	___	___
Pride of workmanship	___	___	___
Joy in learning	___	___	___
Information sharing	___	___	___
Problem solving	___	___	___
Innovation	___	___	___
Partnership development	___	___	___
Customer focus	___	___	___
Quality priority	___	___	___
Fix the system	___	___	___
Do not fix blame	___	___	___
Process emphasis	___	___	___
Cost-benefit analysis	___	___	___
Value-added view	___	___	___
Continuous improvement	___	___	___
Elimination of waste	___	___	___
Reduction of waste	___	___	___
Valuing people	___	___	___

Decision Making

In a democratic society, decisions are usually arrived at by a majority vote. In special situations, a two-thirds vote is required where greater support is needed to make a change. In the TQM group process, the goal for reaching a decision is to arrive at a consensus.

Consensus means that recognition and respect have been given to everyone's ideas, opinions, comments, and suggestions. Every effort has been made to arrive at a decision which accommodates as many ideas as possible. Recognizing that it is virtually impossible to please everyone, the consensus principle is that everyone can at least live with and support the final decision. Consensus requires that the group agree on the point or issue being discussed before it becomes a part of the final decision.

However, for consensus to work, the group itself must be committed to working in a cooperative manner and demonstrate at least a minimal level of trust and respect.

Consensus does not mean everyone is happy with the final decision. Although this is a noble goal, there is simply not enough time, and there are too many conflicting interests and needs to satisfy everyone. What cannot be allowed to happen is one person preventing the group from reaching a decision that the rest can support.

The following guidelines should be used:

- take as much time as possible to discuss all ideas
- show respect and consideration for every idea
- present logical, data-based arguments rather than emotional appeals
- ensure full participation by all
- avoid win-lose situations
- avoid group-think phenomena

Group Decision Charts

We use a series of criteria rating forms to evaluate the relationship between problems identified during the Brainstorming sessions and the team's mission. We also use rating forms to determine the potential of solutions for resolving problems. The rating criteria are determined by the team. Criteria can be treated equally, or they can be weighted relative to each other. We strongly recommend that the criteria be equally rated to each other. In this way, the team is clear in its evaluation process.

The worksheets on the following pages are used by the team during the decision phase:

√ **Problem Selection Process**, criteria defined by the team

√ **Solution Selection Process**, criteria defined by the team

√ **Criteria Rating Process for Problems**, criteria to be defined by the team

The team prioritizes the problem statements according to their ranking and continues to filter the process by using the evaluation sheets on the next pages.

Once the problems have been defined and rated, solutions may be sought and evaluated using the Problem and Solution Selection worksheets.

The following definitions apply to the terms listed in the Problem Selection Worksheet:

Solve: The extent to which the team has the capability to solve the problem.

Importance: The seriousness or urgency of the problem.

Control: The extent to which the team controls the problem and can control the solution.

Criteria Rating Form for Problem Selection

Problem Criteria	Problem Statement	Problem Statement	Problem Statement
Can the team solve the problem?			
Is the problem important?			
Does the team have control of it?			
Has the team agreed that it is a problem the team wants to work on?			
Total Score			

Ranking Key:
3 = To a great extent
2 = To some extent
1 = To a slight extent

Difficulty:	The team's assessment about the difficulty it will encounter while working through the problem to a solution.
Time:	The team's assessment about the amount of time it will need to resolve the problem.
Resources:	The amount of internal and external resources required to solve the problem (people, money, etc.).

The following definitions apply to the terms listed in the Solution Selection Worksheet:

Solve:	The extent to which the team can implement the solution to solve the problem.
Appropriateness:	The degree to which the resources required to implement the solution (money, people, release time, etc.) are available to the team.
Control:	The extent to which the implementation of the solution is within the control of the team.
Acceptability:	The degree to which other administrators, teachers, staff, and stakeholder groups involved in the process will accept the solution and the changes it might impose on the department or organization.
Time:	An assessment of the length of time required to solve the problem.
Resource availability:	The extent to which the resources (money, people, release time, etc.) required to implement the solution are available to the team.

Problem Selection Worksheet

Problem Statements →	Problem Statement	Problem Statement	Problem Statement
Solve 1 2 3 4 5 Low High			
Importance 1 2 3 4 5 Low High			
Control 1 2 3 4 5 Low High			
Difficulty 1 2 3 4 5 Low High			
Time 1 2 3 4 5 Low High			
Resources 1 2 3 4 5 Low High			

Key: In the boxes across the top, write the problem statements the team is considering. Rate each problem statement against the listed criteria by working across each row. The higher the total score, the greater the likelihood that the problem is appropriate for the team to work on.

Solution Selection Worksheet

Solution Statements →	Solution Statement	Solution Statement	Solution Statement
Solve 1 2 3 4 5 Low High			
Appropriateness 1 2 3 4 5 Low High			
Control 1 2 3 4 5 Low High			
Acceptability 1 2 3 4 5 Low High			
Time 1 2 3 4 5 Low High			
Resource Availability 1 2 3 4 5 Low High			

Key: In the boxes across the top, write the solution statements the team is considering. Rate each solution statement against the listed criteria by working across each row. The higher the total score, the greater the likelihood that the solution can be effectively implemented.

How to Use Criteria Rating Forms

To use criteria rating forms:

- decide what factors or criteria are to be considered
- reach agreement on their definitions
- determine what, if any, weights should be assigned
- agree on a point scale to be used
- discuss each "cell" on the form to arrive at a consensus rating

It is best to look at all options (e.g., potential solutions) and rate them on a particular criterion (e.g., the team's ability to control the implementation) at the same time. The team may determine that solution "B" provides the greatest control. Assigning it the highest value then makes it easier to assign ratings to the other portions, relative to solution "B."

Chapter 6

Getting Started as a Facilitator

Getting started as a facilitator is rather easy because it is simply a matter of following procedures; the difficult and delicate part is actually implementing the skills and knowledge.

Steps (New Groups)

1. **Meet with the leader of the group to be facilitated.**

2. **Agree on roles and responsibilities.**

3. **Agree on the manner in which the leader wants facilitation intervention, whether**

    - direct (spontaneous),

    - signal (alert to leader), and/or

    - evaluation (after the meeting).

4. **Review the effective meetings model.**

5. **Develop a strategy for establishing the ground rules to be followed by the group.**

    - "Start from scratch" (have the group develop the ground rules).

    - Prepare a basic list appropriate to the task and group and then have the group make additions or changes (this is much quicker).

- Record the list and review it prior to each meeting. It is preferable to display the list where it is always visible—a flip chart page posted on the wall is sufficient.

- Use a contract.

6. **Decide on how the recording will be done.** One option is to have an officially designated recorder either from outside the group or a member of the group. The problem with a member of the group recording is that it is more difficult for the recorder to participate fully. The other option is to have the facilitator fulfill this role by using a flip chart, and this does make it easier to facilitate the meeting flow. "He who controls the flip chart controls the meeting." In fact, this is an excellent way for the facilitating interventions to take place without threatening the authority of the leader, and it is natural to do.

7. **Make sure that any needed equipment and materials are in place and that the room is set up in an appropriate manner.** The worse thing to have happen is to delay the meeting or any part of the meeting because the required equipment or materials are not available.

8. **Start the meeting on time (job of the leader).**

9. **Begin with introductions (if needed).**

10. **The leader should introduce the facilitator and give an explanation of how the facilitator will function.** The facilitator should then give a personal introduction.

11. **Go over the agenda.** This should include time factors, break times, etc.

12. **Ask the group if there are any questions and answer them.**

13. **Review and complete the Team Formation Guideline form (see Appendix).**

14. **Decide on the ground rules.**

15. **DISCUSS HOW THE GROUP WILL EVALUATE ITS EFFECTIVENESS.**

- Formal (survey form—if so, how often?)

- Successful accomplishment of the task, goal, or agenda

- Other method or combination of the above

16. **Begin the meeting.**

17. **EVALUATE THE MEETING WITH THE LEADER.**

Steps (Functioning Groups)

1. **Review with the leader all of the above to determine whether anything needs to be done that should have been done.**

2. **Ask the leader how the group is doing and whether there are any problems.** If there are, develop a strategy for dealing with the problems.

3. **Determine whether the leader wants to conduct a formal team effectiveness survey.**

4. **If a survey is used, analyze the results and provide feedback to the group.** The leader should determine who will do this.

5. **Determine whether any changes need to be made. Get agreement.**

6. **Get back on the task.**

Sample Meeting Ground Rules

The following meeting ground rules were developed by teams to improve team performance and effectiveness.

√ **Leave positions at the door.**

- Position does not have its privileges on the team.

√ **Respect others.**

- Listen—everyone owns part of the process and everyone needs to be heard.

√ **Use "I" statements.**

- Does not work: "But you said…"

- Does work: "I understood you to say that…"

√ **Think about the meeting beforehand.**

- Team activities may require change.

- Be prepared to fully participate at team meetings.

- Review team ground rules.

- Contribute to the success of the meeting.

√ **Review the agenda at the beginning of every meeting.**

- Agree on break and lunch times.

- New items can only be added to the agenda if consensus is received from the team.

√ **Start and end the meeting on time.**

- Don't penalize team members for arriving on time.

- Reinforce the importance of the team process.

√ **Bring the team to unity; obtain team consensus.**

- Consensus: "I will agree to support the decision."

- Agreement: "I agree with the decision."

- Conflict: "I cannot support or agree with the decision."

√ **Create team meeting minutes.**

- Use flip chart notes and review them at the end of the meeting for accuracy.

- Send meeting minutes to everyone before the next meeting

- Review team assignments at the end of the meeting.

√ **Evaluate the team meeting.**

- Were the desired results achieved?
- Did everyone contribute to the meeting?
- Were team processes followed?
- Was the facilitator effective?

√ **Stay through the hard parts.**

- Members must resolve all conflicts before the end of the meeting.
- The meeting cannot end on a "sour" note.
- No one should leave early.
- The team must reach consensus.

Exercise

Use the space below to develop team meeting ground rules for your organization. Remember, team meeting ground rules must be supported and approved by all team members.

_____

_____

_____

_____

_____

_____

Chapter 7

Conducting Effective Team Meetings

There are numerous publications and videos on ways to conduct effective meetings. What follows is a quick guide to emphasize the critical need to have an effective procedure for meetings. The cost of meetings and the time expended for meetings are considerable. When meetings are unproductive or unnecessary, it is wasteful.

Typically, organizations spend 7 to 15 percent of their personnel budget, 35 percent of middle management's time, and 60 percent of top management's time in meetings—representing more than 11 million meetings a day.

Therefore, one of the first things to do is to determine how many meetings take place on a regular basis throughout the organization and how many others take place not on a regularly scheduled basis. Determine from each group, through written surveys, how effective the meetings are and whether they are even necessary.

Also, have each group determine the cost of the meeting (multiply the number of people × the average cost per hour × the length of the meeting, as well as preparation time). Add up the weekly, monthly, and yearly cost. Do it by department and functional area, as well as system-wide.

This data will provide a benchmark to compare the savings by having more effective and perhaps fewer meetings. By having more effective and fewer meetings, human resource time can be freed up for the continuous improvement process.

Useful Models, Tools, and Techniques

→ Surveys

→ Checksheets

→ Cost-Benefit Analysis

→ Conflict Management

→ Problem Solving

Effective Meeting Model

As discussed in the previous chapter, every team should have ground rules which will govern the team's activities and meetings. One of the major causes of failure in conducting effective meetings is the inability of people in a team to build trust among themselves. Quality teams should also build partnerships with all the stakeholders of the process. As the level of trust increases among the team members and between the team and the stakeholders, the team will create better learning and working environments. The following are successful hints for managing team meetings:

1. Code of Conduct

One of the first organizational items the team should discuss is the code of conduct. Individual members should agree to abide by the code of conduct. Members who cannot abide by the code of conduct should be excused from the team.

2. Meeting Location

Schedule a convenient meeting location. Make sure that the location is reserved well in advance of the meeting. Establish a date and time for the meeting that is convenient for every team member.

3. Meeting Attendance

Meeting attendance is mandatory, but there may be a legitimate reason why someone cannot attend a meeting. When someone cannot attend a meeting, he/she must contact the team leader and explain the reason. When someone misses a meeting, he/she agrees to accept any task assigned by the team. If someone consistently misses meetings, he/she should be asked to resign from the team, and a replacement should be selected.

4. Promptness and Attention

The meeting should start and end on time. The participants should give the meeting their full attention. They should not be trying to complete another task while at the meeting. Phone calls should not be allowed except for emergencies. Distractions should be kept to a minimum. Break periods should be established at the beginning of the meeting.

5. Agenda and Minutes

The team should agree on the next meeting's agenda prior to the conclusion of the meeting. The team recorder should keep minutes of the meeting. The minutes of the previous meeting and the agenda for the next meeting should be sent to the team members in advance of the meeting. The minutes should reflect the tasks that have been assigned to individual members. The team recorder should keep a central file of the minutes of the team meetings. The central file should also contain the project reports submitted by team members as they conclude their projects.

6. Roles and Responsibilities

Generally, the team will assign specific tasks to team members. However, there may be regular duties that the team may assign to one or two members. These tasks should be assigned to the members at the initial meetings.

7. Team Skill Assessment

Early in the problem-solving process, the team should determine the areas of expertise for each member of the team. This will enable the team to identify the external resource requirements that may be needed to solve the problem.

8. Resource Requirements

This is a major issue that is often overlooked. Very early in the problem-solving process, we recommend that the team develop a list of resources that will be needed to resolve the problem.

9. Project Schedule

Every team should establish a project schedule prior to beginning work. This will enable the team members to schedule the team meetings in their calendars. It will help management to understand the process the team will use to the solve the problem. It provides the team with specific milestones and objectives by which the team can measure its progress. Developing a project schedule helps the team to be efficient and eliminate wasted effort.

10. Communication

The team should establish a process by which the team can communicate its work to others. Team communications should summarize the team's activities. Whenever possible, use existing communication processes to keep others informed.

Characteristics of Effective Teams

The following are characteristics of effective team meetings:

1. Promptness

Meetings should start and end on time. Members should come to the meeting prepared to participate in the team activities.

2. Participation

Every team member is expected to participate in all of the team's activities.

3. Basic Conversational Courtesies

Everyone should listen attentively and respectfully to others. Team members should not interrupt each other. Only one conversation at a time should occur, and members should not pass notes to each other while someone is trying to speak.

4. Agenda and Minutes

The team recorder is responsible for keeping minutes of the meetings. The agenda should be published for every meeting and reviewed at the beginning of every meeting.

5. Breaks

At the beginning of every meeting, the team should decide when and how long breaks will be. The team should break at the designated time, and the team members should resume activities at the agreed time.

6. Assignments

Much of a team's work is done between meetings. When members are assigned responsibilities, it is important to complete tasks on time. If a team member is unable to complete the task according to the schedule established by the team, he/she should bring the matter to the attention of the entire team. The team may have to assign additional resources to the task or assign the task to another team member.

7. Discussions

Everyone is encouraged to participate in the team's discussion. Members should not attack one another for their thoughts. It is okay to attack the idea, but it is not okay to attack the person. People should be made to feel that their opinions are valued.

8. Next Meeting Agenda

At the conclusion of every meeting, the team should establish an agenda for the next meeting. The team should answer the question, "What must we accomplish next to complete the project on time?" The development of a team agenda also enhances the capability of the team. The agenda is owned by the team and not by a single individual.

9. Meeting Evaluation

Evaluation is the most important and difficult activity the team will undertake. Self-critiquing is a team's main source of feedback and the only way to avoid letting problems go unnoticed.

10. Meeting Close

Always end the meeting on a friendly note. No one should leave the meeting feeling unappreciated. Sometimes discussions become very heated and painful. It is the facilitator's responsibility to ensure that harmony exists at the end of the meeting.

Principles of Effective Team Management

The following are principles for effective team management:

1. Meeting Management

The facilitator should ensure that the team's activities are governed by the principles of quality.

2. Team Discipline

This is a very difficult issue for the team to address. It is one of the main reasons we strongly recommend that an external facilitator initially be used to manage the team's activities. The major areas of concern are attendance, completing tasks on time, uneven work load, lack of participation at meetings, and lack of results. An external facilitator is objective and does not have a vested interest in the process or the people.

3. Maintaining Focus

In the previous section, we strongly recommended that the team establish a project schedule. An important check for the team is to review the progress the team makes against the project schedule. This will enable the team to recognize problems that the team is having in meeting its schedule and to take corrective action. It is also an opportunity for the facilitator to reinforce the importance of the task and the need to maintain a schedule.

Effective Meeting Exercise

1. Review

Design a survey form to determine what people in the organization think about meetings. Consider the following categories for the survey:

*   benefits

*   disadvantages

*   suggestions for improving

*   reducing the number of meetings

*   reducing the amount of time allocated for meetings

*   effectiveness of meetings

*   substitution for meetings

2. Cost of Meetings

Identify the last six meetings, particularly those held on a regular basis.

Use the following data to determine the cost of each meeting:

- Hourly pay for the specific individuals who attended the meeting.

- Multiply the hourly pay for each individual × the total time allocated to the meeting.

- Calculate the cost of secretarial time for notices, phone calls, discussions, etc.

- Calculate the cost of any individuals involved in preparing for the meeting, such as developing the agenda, preparing materials, presentations, etc.

- Calculate the cost of any materials used.

- Calculate the cost of any refreshments provided.

- Calculate the cost of the room used unless it would have been occupied for other use (if it was held in someone's office, there would be no real added cost).

- Total all costs. Determine whether the cost justified the results or outcomes or whether the same outcome or results could have been achieved with lower cost.

3. Evaluate

Have the leader involved with a facilitator evaluate the prior meeting (if there was one) and the next three meetings by using a checklist or form. The following is an example of an evaluation form we use to assess the effectiveness of the facilitator and technical assistance we provide our customers.

Facilitation Evaluation

Please rank the facilitator's contribution to the team meeting according to the following criteria:

	Excellent	Very Good	Satisfactory	Poor	Unsatisfactory
1. Did the facilitator/technical advisor have an accurate perception of what you needed and wanted?	☐	☐	☐	☐	☐
2. Were the objectives of the session clearly identified?	☐	☐	☐	☐	☐
3. Were the outcomes of the session clearly identified at the beginning of the session?	☐	☐	☐	☐	☐
4. Were the outcomes agreed to by the team?	☐	☐	☐	☐	☐
5. Did the outcomes match your expectations for the session?	☐	☐	☐	☐	☐
6. Was the sequencing of the session and material logical?	☐	☐	☐	☐	☐
7. Were you provided with the opportunity to discuss your ideas at the meeting?	☐	☐	☐	☐	☐
8. Did you get satisfactory answers to your questions?	☐	☐	☐	☐	☐
9. Did the facilitator/technical advisor use only language that is very clear to you?	☐	☐	☐	☐	☐
10. Do you feel that each step in the problem-solving process was satisfactorily explained to you before the next step was started?	☐	☐	☐	☐	☐
11. Was the facilitator/technical advisor easy to listen to and follow?	☐	☐	☐	☐	☐
12. Did the meeting achieve the stated outcomes?	☐	☐	☐	☐	☐
13. Are you satisfied with this session?	☐	☐	☐	☐	☐
14. Did the facilitator contribute to the success of the meeting?	☐	☐	☐	☐	☐
15. Would you recommend this facilitator for future use?	☐	☐	☐	☐	☐

Conflict Management

It has been said that there are only two things predictable about life—death and taxes; however, a third predictable life pattern is "conflict."

Conflict has been defined in many ways, but in simple terms, it is a disagreement caused by dissimilar points of view or unacceptable behavior between two or more individuals or groups. In its most destructive form, it has caused war and death, but in its constructive form, it can be an opportunity to negotiate and resolve differences.

The "dissimilar points of view" occur for many reasons, but some principal causes involve:

- differences in personalities
- values and beliefs
- performance evaluation
- work assignments
- roles and responsibilities
- criticism
- misunderstanding
- frustration
- work overload
- problem solving
- competition
- policies, rules, and regulations
- poor communication
- use of power

The importance of having a conflict management model cannot be stressed too strongly because unless conflict is handled constructively, it will be costly in terms of time, energy, wasted resources, and poor quality outputs.

It has been estimated that managers spend 24 percent of their work time in handling and resolving conflict.

Useful Tools and Techniques for Resolving Conflict

The following tools and techniques can be used to resolve team conflicts. As previously stated, it is important to resolve conflict before the end of the team meeting.

Useful Tools and Techniques

→ **Brainstorming and Brainwriting**

→ **Cause-Effect Diagrams**

→ **Surveys and Interviews**

Using the Conflict Management Matrix, list the areas of conflict in the column on the left. In the columns across the top, list those organizations or individuals that are impacted by the areas of conflict. Rank the conflict according to the degree of impact on the organization or individual. Resolve the issues with the highest ranking first. An example follows of how Galileo used the Conflict Management Matrix in a college to resolve and eliminate areas of conflict.

Conflict Management Matrix

Departments, Organizations, Individuals

Areas of Conflict					

Instructions: Rank the degree of conflict, on a scale from 1 to 5 with 5 being the highest, for each item listed in the Areas of Conflict column. Focus on reducing or eliminating those items with a high degree of conflict. Your program will be more successful if you can minimize the conflict of individuals, groups, or departments.

Areas of Conflict

Departments, Organizations, Individuals

Areas of Conflict	Senior Management Team	Program Teams	Teaching Staff	All Other Departments
Lack of clear understanding of roles and responsibilities	High	High	High	High
Lack of clear understanding of policies and procedures	High	High	Medium	High
College reorganization	Medium	High	High	High
Lack of understanding of college's future direction	Medium	High	High	High
Financial concerns	Medium	High	High	High
College management team	Low	Medium	High	Low
Galileo project	Low	High	High	Low
Incorporation	Medium	High	High	High
Constant change direction	High	High	High	High
Lack of concern for the individual employee	High	High	High	High
Lack of communication between senior management and staff	Medium	High	High	High

Conflict Management Model

1. Prevention

The most effective way conflict can be managed constructively is to anticipate when and where it can occur and then plan appropriate intervention strategies. There are two procedures necessary to anticipate conflict.

2. Search for patterns of conflict

a. List all of the persistent and/or major conflicts which have occurred over a given period of time (one year, if possible). This task should be accomplished by a team rather than an individual.

b. Separate the list to determine who (individuals, groups, departments, etc.) was involved in the conflict.

c. Categorize the cause of each conflict. To get started, use the principal causes list on page 68.

d. Next to each conflict, indicate the functional area where it occurred (central office, building, department, etc.).

e. Indicate whether or not the conflict was resolved satisfactorily.

f. Study the data to find the patterns of conflict.

g. Examine the data to determine whether the conflict could have been prevented and, if so, how—what could have been done?

h. Develop and implement strategies or procedures that could prevent or minimize future conflicts.

3. **Anticipate when and where conflict will occur**

    a. Before implementing any changes in procedures, policies, personnel, etc., develop a matrix indicating, on the vertical axis, what the change is and, on the horizontal axis, who will be impacted.

    b. Determine what needs to be done to prevent or minimize conflict from occurring. For example, training and/or education may be required for the individuals involved; communicating effectively with those involved may be necessary to ensure that everyone understands what the changes are and why the changes are being made.

4. **Understanding**

People have different conflict management styles. Helping them to know their style can be helpful in handling conflict more constructively. There are a variety of survey instruments available that can be used to assess one's style.

    a. **Choose an instrument to be administered to administrative personnel first because the administrators must model constructive conflict resolution behavior.**

    b. **Conduct a workshop to discuss conflict management styles and what styles are most effective in conflict prevention and resolution.**

5. **Negotiation**

Although conflict can be minimized, it cannot be eliminated. Therefore, conflict resolution becomes a key factor in keeping harmony, developing cooperation, and creating teamwork.

    a. **Use a text on negotiating successfully (there are many available) if there is in-house expertise available; otherwise, it would be wise to choose a competent consultant.**

b. **Conduct a workshop series for administrators. One workshop will not be sufficient since practice is required along with follow-up to reinforce skills and competencies.**

Although this may seem time consuming and less of a priority in light of other administrative demands, reducing the stress and energy involved in daily conflict will free up time and energy needed to meet the demands and requirements of a quality organization.

Conflict Management Exercise

In the space below, complete the following activities.

1. Prevention—Patterns of Conflict
Search for patterns of conflict by following the steps outlined on pages 72–74.

2. Prevention—Review
Recreate the events of a recent conflict situation where some change process was implemented.

Make a matrix and complete it as if the change and conflict had not yet occurred. Would the matrix have helped to identify the conflict? Would it then have been possible to either reduce the intensity of the conflict or perhaps eliminate it entirely?

3. Prevention—Anticipate
Prior to the next change implementation, use the matrix to anticipate conflict.

4. Evaluate

Evaluate the use of the matrix after it has been used several times.

Communication Model

When people are asked to take on a task, they want to know that what they do will make a difference, and, furthermore, they want to be given some recognition for their efforts.

One very effective way to accomplish both needs is the effective communication of the team's activities and results. Usually, a management presentation or a formal report is given to management prior to the conclusion of team activities. Additionally, the team generally prepares a briefing package for dissemination throughout the organization. Communication is the only way to build broad support and consensus for the team's activities. Some of the tools used to communicate the team's activities are as follows:

Useful Models, Tools, and Techniques

→ Cost-Benefit Analysis

→ Cause-Effect Diagrams

→ Checksheets

→ Force-Field Analysis

→ Histograms

→ Pareto Charts

→ Surveys and Interviews

Communication Techniques

People communicate in a variety of ways. Communication styles can be categorized as:

- action focused
- relationship focused
- thinking focused

Teams that practice action-focused communication concentrate on the results of the team's activities. As a result, teams that practice this type of communication may seem to be detached, independent, and competitive. Their communication must be clear-cut and related to actions.

Teams that utilize the action-focused techniques tend to concentrate on the present. They are swift, efficient, and to the point. They show little concern for the past or the future. The team likes to work quickly, and members have difficulty working with other teams or individuals who do not have the same work ethic.

Action-focused teams make their own decisions. They seek power. They are quick to say what they think. When obstacles block their path, they do whatever it takes to work through them.

Relationship-focused communication is based on the concept of cooperation, collaboration, and partnership development. Teams that practice relationship-focused communication lend freshness and warmth to any situation. Team members regard the feelings of others. They look for the "why" in the action of others.

Power over others does not motivate relationship-focused teams. Being accepted by others is important. These teams also like to get things done. However, the team uses understanding, respect, and cooperation to get things done.

To action-focused teams, relationship-focused teams appear to be undisciplined. They are viewed as wasting time. They seem to move slowly. Socializing may seem to be more important than accomplishing the task. Action-focused teams find it difficult to work with relationship-focused teams because they view relationship-focused teams as being conforming, inefficient, pliable, and overly agreeable.

Thinking-focused communication is best described as a "show-me" attitude. Teams that practice thinking-focused communication are suspicious of power or leverage. These teams have a predictable communication pattern.

Thinking-oriented teams avoid flashiness; they move slowly and in a disciplined, deliberate way. Team members look at information in a calm, common sense way, focusing on the past to gain purpose, meaning, and direction for the future.

Communication Exercise

Review the material previously discussed and complete the following action items to develop a communication model for your organization.

1. Review

Identify a recent situation where a "formal" presentation was made to request something and either it was denied or it took a great deal of discussion.

Analyze the circumstances and determine the communication style.

2. Model

Use the information to develop an effective communication model for the team in your organization.

3. Evaluate

Try it out. Did it work?

Developing Good Communication Skills

The following are techniques that you can employ to develop good listening skills.

Listening Pointers

◊ Stop talking—listen

◊ Empathize with the other person

◊ Ask questions

◊ Look at the other person

◊ Leave emotions behind

◊ React to ideas—not the person

◊ Don't antagonize

◊ Avoid assumptions

◊ Listen to what is not said

◊ Don't argue mentally

◊ Get rid of distractions

◊ Concentrate on the main points

◊ Paraphrase (Is this what I hear you saying...)

◊ Don't interrupt

◊ Don't assume what the person is saying

◊ Offer advice only when asked

Observing

◊   Unspoken messages

◊   What to do next

◊   What does it mean?

Positive Reinforcement

◊   Nodding head affirmatively

◊   Voicing agreement

◊   Smiling

◊   Repeating the statement

◊   Requesting more information

◊   Listening carefully

◊   Asking the group members to comment

◊   Asking to repeat statements

◊   Touching the person

◊   Giving a "token"

◊   Saying "thank you"

◊   Complementing the remark

◊   Writing it down (flip chart)

Observing

The purpose of paying close attention to your group or team members is to observe what they are saying—and not saying.

About 50 percent of all communications is conveyed non-verbally. A facilitator must watch for the unspoken messages that accompany what is said. Utilizing observation skills can help to assess the effectiveness of the group process and how well information is being received. Based on these observations, adjustments can be made in questions, a new activity or procedure, a break, etc.

Observations collected over time can help to build a file of mental or written notes about what is effective in terms of interventions.

Look for these body language signs:

- **frowning**
- **nodding**
- **yawning**
- **shuffling feet**
- **tapping a pencil/pen**
- **looking at you or away**
- **smiling**
- **leaning back or forward**

Chapter 8

Problem-Solving Tools and Techniques

Problem-Solving Model

The following six steps provide a foundation for identifying and solving problems.

1. Identify and Select the Problem

   a. Determine what is causing the problem or where a gap exists

   b. Develop a problem statement

   c. Decide what the desired state should be in measurable terms

   d. Determine what needs to change

Useful Models, Tools, and Techniques

→ Affinity Diagrams

→ Brainstorming/Brainwriting

→ Flowcharts

→ Histograms

→ Interrelationship Diagraphs

→ Matrices

→ Management Information System

→ Pareto Charts

Recommended Tools to Identify and Select the Problem

The Affinity Diagram and Interrelationship Diagraph are excellent tools to use to get a group of people to identify and select the problem. The Affinity Diagram is used to generate ideas relative to identifying the problem statement. The Interrelationship Diagraph takes the problem statement and maps out the logical links among related items.

---

2. Analyze the Problem

   a. Determine what is preventing the desired state from being achieved

   b. Identify key causes and reasons

   c. Document and rank key causes and reasons

   d. Identify root cause(s)

Useful Models, Tools, and Techniques

→ Cause-Effect Diagrams

→ Brainstorming/Brainwriting

→ List Reduction

→ Mind Mapping

→ Tree Diagrams

Recommended Tool to Analyze the Problem

The Tree Diagram maps out in increasing detail the full range of paths and tasks. This tool enables you to analyze the problem in greater detail.

3. Generate Potential Solutions

   a. Develop and clarify potential solutions

   b. Rank order the solutions

   c. Determine how the change(s) can be made

Useful Models, Tools, and Techniques

→ **Force-Field Analysis**

→ **Cost-Benefit Analysis**

→ **Nominal Group Technique**

→ **Prioritization Matrices**

Recommended Tool to Determine Potential Solutions

The Prioritization Matrix is used to prioritize tasks, issues, and opportunities to determine.

4. Select and Plan the Solution

   a. Focus on one solution to reach the desired state

   b. Determine the best way to make the change(s)

   c. Develop criteria for evaluation

d. Develop an implementation plan

e. Verify that there is consensus for the plan

Useful Models, Tools, and Techniques

→ List Reduction

→ Management Presentation

→ PERT Charts

→ Process Decision Program Chart (PDPC)

Recommended Tool to Select and Plan the Solution

The PDPC tool is a method which maps out conceivable events and contingencies that can occur in any implementation plan. It identifies feasible countermeasures in response to these problems.

5. Implement the Solution

a. Execute the plan

b. Monitor the implementation

c. Track the measurement components

d. Develop a contingency plan

Useful Models, Tools, and Techniques

→ Checksheets

→ Histograms

→ Pareto Charts

→ Activity Network Diagrams

Recommended Tool to Implement the Solution

The Activity Network Diagram is used to plan the most appropriate schedule for the completion of any complex task and all of its related sub-tasks. It projects likely completion time and monitors all sub-tasks for adherence to the necessary schedule.

6. **Evaluate the Solution Implementation**

    a. Check critical success factors

    b. Check measurements

    c. Measure the effectiveness of the solution

    d. Verify that the problem is solved

Useful Models, Tools, and Techniques

→ **Interviews**
→ **Surveys**
→ **PERT**
→ **Flowcharts**
→ **Decision Charts**

Problem-Solving Exercise

Complete the following activities to implement the problem-solving model in your organization.

1. Review

Review a problem that was addressed recently when no problem-solving model was used.

Using the six-step model, determine whether following the model would have helped to solve the problem more effectively, quickly, easily, etc.

2. Select a Problem

Select a problem that needs to be addressed and use the six-step model. This should be done for at least two or three other problems.

3. Evaluate

Evaluate whether the model was useful in solving the problem(s).

_____

_____

_____

_____

_____

_____

_____

_____

_____

_____

_____

Chapter 9

Team Formation

The problem with the formation of any team, committee, or task force to solve a problem or address an issue is that three fundamental mistakes are made:

1. There is never sufficient clarity up front as to exactly what is expected of the group.

2. It is assumed that the members of a group have the necessary knowledge and skills to be effective.

3. There is an assumption made that assembling a group or forming a committee makes a team.

It takes time to make a team! Just think about the time and practice required to develop a professional athletic team even though every member:

*   is highly skilled

*   is experienced

*   knows his/her exact role

*   knows his/her exact responsibility

*   understands the rules of the game

Although the staff may be highly educated, they have not been trained in team skills or team building; in addition, roles, responsibilities, and rules are clouded with many shades of gray. Therefore, more time and practice are required to develop a team if effective problem solving is to occur.

Other factors too often overlooked in advance are the personal and professional skills and knowledge needed on the team if the team is to be really successful. What usually happens is that a decision is made as to who should be on the "committee" without regard to the skills and knowledge required.

The nature of the problem to be solved or issued to be addressed should dictate what is required. For example, if a new program is being considered, some skills and knowledge required would include:

• research

• budgeting and financial analysis

• testing

• planning and organizing

• analyzing

The model that follows assumes that members of a team have had training in team building and team skills. If such is not the case, training should be done first, and a facilitator should be assigned to the group.

Useful Models, Tools, and Techniques

➜ Brainstorming

➜ Checksheets

➜ Matrices

➜ Nominal Group Technique

➜ Surveys and Interviews

Selection of Team Members

There are several methods that you can use to form your team. The method we use is the *stakeholder process*. In the stakeholder process, we identify all of the people or organizations that are customers of or suppliers to the process under review.

Membership in a quality team is usually voluntary. However, members can be appointed to serve on task teams. They can be appointed to serve on the team because of their position of employment; policy may dictate representation on the team. The members selected to serve on the team must have an interest in solving the problem, and their field of expertise may necessitate their participation in task teams.

However selected, task team members cannot be uninterested individuals who are forced to serve on a team. They must want to help solve the problem. If they have no interest in solving the problem, they will not contribute to the activities of the team, and, more importantly, they will not support the team's recommendation. This makes it extremely difficult for the team to gain broad support for its solution.

The following guidelines can be followed by the organization to create task teams.

Team Formation Model

1. Clearly state in writing the problem or issue to be studied.

2. Indicate the desired result(s):

a. measurable

b. non-measurable

c. time frame

3. State the relationship to one or more of the following:

a. vision

b. mission

c. beliefs/values

d. mandates

4. Clarify the constraints:

a. budget   g. services

b. staff   h. information

c. time   i. administration

d. technology   j. data access

e. equipment   k. policy/regulations

f. programs   l. legal

5. List the assumptions which have been made.

6. Identify the team members (names and/or positions) if they have been assigned or indicate how volunteers will be solicited.

7. Decide how a chairperson will be selected and what responsibility has been assigned.

8. Indicate whether minutes are to be taken and how, when, and to whom they will be distributed.

9. Clearly state what compensation (time or money), if any, will be given.

10. Indicate the meeting times.

11. State what resources, if any, will be provided.

12. Decide how decisions will be made.

13. Indicate the meeting guidelines to be followed.

14. Indicate how reporting will be done, who will receive the report(s), and when.

15. Make very clear exactly what authority the group has been given.

Team Formation Exercise

In the space below, complete the action items to develop a team formation model for your organization.

1. Review

Identify a committee, task force, or team and use the fifteen guidelines to determine whether all the steps have been followed.

2. Model

If a team, committee, or task force is to be formed, follow the fifteen guidelines.

3. Evaluate

Survey the teams in #1 and #2 above to determine whether they believe the guidelines are or were helpful.

_____

_____

_____

_____

Chapter 10

Creating a Customer Focus

An organization can collect and utilize data, make effective use of tools and techniques, have a clear understanding of how the work gets done, and implement all the management models and still not be considered an effective organization unless there is a focus on the *customer*. Many businesses have learned this critical principle of quality the hard way and, unfortunately, many others still do not understand the need for a customer focus.

If it is difficult for business to understand, it is even more difficult for educators and government professionals to understand and appreciate. Government service is a human service enterprise and, by and large, it has a *captive* customer base. Complicating the issue of customer focus in schools is trying to get agreement as to who the schools' customers are: students, parents, the entire community, etc.

Once the *internal and external* customers are identified, the next difficult task is trying to meet customer requirements and expectations (CREs). In manufacturing a product or providing a service, it is much easier to determine what the customer requires and expects. In government, customer requirements and expectations may be difficult to meet because resources are limited and there is no real consensus as to the specific quality outputs desired and needed.

Nevertheless, if customers are not satisfied, and all such surveys indicate that this is the reality, resources will be difficult to obtain and support will erode; worse yet, changes will be legislated by those who have limited knowledge and understanding of the complexities involved in education.

Who is considered to be a customer? The broadest definition is "the person, department or organizational unit that next receives the value-added product, service or client." This definition means that at some point in the process, everyone is considered and treated as a customer. In a human enterprise system, such as education, this is a powerful concept because it addresses the reason why cooperation and teamwork are so essential.

However, internal customer expectations and requirements can only be addressed within the context of meeting the ultimate customer requirements and expectations.

Useful Models, Tools, and Techniques

→ **Surveys and Interviews**
→ **Affinity Diagrams**
→ **Interrelationship Diagraphs**
→ **Brainstorming**

The following steps can be used by teams to identify the primary, secondary, and ultimate customers.

Customer Focus Model

1. Identify every group that benefits from the enterprise.

2. Identify the internal groups vs. the external groups.

3. Agree on who (is) (are) the ultimate or primary customer(s).

4. Survey and/or interview the ultimate customer(s) to determine their requirements and expectations.

5. Determine which requirements and expectations are legitimate and reasonable.

6. Analyze the system capability to meet the expectations and requirements.

7. If the system does not have the capability, notify the customers and negotiate what needs to be done.

8. Develop a plan of action with measurable goals and objectives.

9. Communicate the plan to the rest of the enterprise and other stakeholders.

10. Survey and/or interview the internal customers to determine their requirements and expectations.

11. Determine which are legitimate and reasonable within the context of the requirements and expectations of the external customer(s).

12. Analyze the system capability to meet the expectations and requirements.

13. Communicate to the staff what the enterprise can do and how it intends to meet their expectations and requirements.

14. Develop a procedure to ensure that the plans for the external and internal customers are implemented and monitored for success.

Customer Focus Exercise

Review the material previously discussed to complete the following activities. This exercise will help you develop a customer focus model for your organization.

1. Review

Determine who has been treated as the "customers."

Agreement or consensus does not have to be reached. The purpose of the review is to determine what differences exist about who has been considered the "customer."

2. Model

Go through the fourteen guidelines to identify customers—this, of course, will be a long process.

To simplify the exercise, use the group you are in and follow the appropriate guidelines so that there is an understanding of what needs to take place.

Consensus must be reached.

3. Evaluate

Evaluate the model and the effectiveness of identifying and communicating with "customers."

_____

_____

_____

_____

_____

Chapter 11

Basic Tools of Quality

The tools outlined in this chapter are used by facilitators at team meetings to improve the effectiveness, efficiency, and productivity of team meetings.

Brainstorming

Brainstorming is a technique to generate as many ideas or options as possible through group participation in order to achieve quality results. It uses the collective thinking power of a group by generating creative, spontaneous, and numerous ideas in response to a problem or issue, and it prevents one person from monopolizing the group. It is important to understand that brainstorming does not solve a problem; rather, it simply provides a variety of solutions to be considered. The suggestions or options must then be evaluated to determine what may work best.

Brainstorming is not new; it was fully developed and extensively used by the ancient Greeks. Today, it is used quite commonly in the business sector; in fact, Walt Disney helped to popularize it as far back as 1928—it was called *storyboarding*. However, its effectiveness did come into question in the 1950s when some research concluded people were more effective working alone to solve problems. This led to more extensive and conclusive research indicating that, in most cases, group brain-storming was more effective than individual effort.

Significant advantages of brainstorming are that it is easy to learn, it does not take much time, and those involved in the process are all equal in making suggestions, with no constraints to be considered initially.

In most situations, the brainstorming activity is done verbally. If, however, there is a sensitive issue involved, it should be done with brainwriting so that no one feels "on the spot" or reluctant to make suggestions.

Brainwriting is the same as brainstorming, except the suggestions are put on cards or "Post-Its" and passed to the recorder, who posts them on a flip chart. As a result, no one knows who said what.

The "storming" technique is used when a given situation, problem, or issue cannot be solved or addressed through some formula, policy, regulation, or the system itself. In addition, it is used when the efforts of an individual or very small group have not achieved or cannot achieve successful results.

All that is needed is a flip chart, a trained facilitator, "Post-Its" or cards, masking tape, felt markers, adequate space, and the group.

There are some variations of the brainstorming process but, in essence, the basic steps are the same.

Brainstorming Model

The following thirteen-step process is used to implement an effective brainstorming or brainwriting session.

1. Identify the problem to be solved or the issue to be addressed and write it on a flip chart; then, post it on a wall in clear view of the group.

2. Make clear or decide as a group (whichever is appropriate) what the expectation is from the brainstorming session.

3. Have a trained facilitator lead the group.

4. Select a recorder (sometimes the facilitator can function in this role).

5. Start clockwise and have everyone in turn offer one idea at a time. The recorder writes the exact statement on the flip chart. It is preferable not to use "Post-Its" or cards (unless the group is implementing the brainwriting technique) because the group will not be able to read the ideas and suggestions. As each page is filled, place it on a wall in clear view. In this way, the ideas are constantly in view and help generate other ideas.

6. Keep it fast paced—as more ideas are presented, synergy begins to build.

7. No one is allowed to comment **in any way** on the suggestions or ideas.

8. If a group member has nothing to suggest, the person can "pass."

9. After everyone has had a turn, begin the process again in order to generate more ideas.

10. When it becomes obvious that the group has exhausted ideas and suggestions, stop this phase of the process.

11. The next phase is to have group members ask for clarification of any ideas or suggestions.

12. Once ideas are clarified, they are categorized under main headings or topics, and those that are similar are combined.

13. The ideas are evaluated and the group chooses the best solution. The Nominal Group Technique (page 123) is very useful if there is difficulty deciding on the best solution.

Brainstorming Exercise

Review the following material and complete the action steps listed below. This will provide you with a structured process for using the brainstorming and brainwriting techniques.

1. Review

Are there groups, teams, etc. using brainstorming?

_____

_____

_____

_____

Do they follow the model or do they have variations?

_____

_____

_____

_____

What do they think of brainstorming?

_____

_____

_____

_____

What problems have they had in the brainstorming process?

_____

_____

_____

2. Practice

Identify a problem or an issue that needs to be addressed.

PROBLEM _____

For example, why should TQM be implemented (advantages and disadvantages)?

Follow the thirteen-step process and conduct a brainstorming activity.

3. Evaluate

Evaluate the practice activity. If it did not go well, try another problem.

Cause-Effect Diagrams

The Cause-Effect Diagram is also known as the Fishbone (it looks like a skeleton of a fish) or Ishikawa Diagram (named after Dr. Kaoru Ishikawa, who developed this quality tool in the early 1950s). It is a graphic, structured, and systematic way of looking at an effect (the problem) and the causes, conditions, or relationships that bring about (or could bring about) the "effect."

The purpose of using this tool is to generate as many ideas as possible concerning the conditions or causes related to the "effect." Generating the ideas is the same as the brainstorming process. What is different is that each main bone of the diagram is a category within which the brainstorming ideas are included.

The problem (actual or potential) is written in the main box which is at the end of the arrow or main horizontal bone. One category is then placed in a box at the top of a secondary bone connected on an upward slant to the main bone (arrow).

Typically four secondary bones are used along with four categories (one for each bone). The usual categories are manpower, machinery, methods, and materials; sometimes measurement is a fifth bone. However, no more than four to eight categories (bones) should be used, but the categories can be different.

Some examples of fishbone categories in a school situation follow:

√ **Decision-Making Problem**

- realistic
- readiness
- resources
- receptiveness

√ **Human Relations Problem**

- professional
- physical
- personal
- psychological

√ **Group Process Problem**

- skills
- support
- system
- subversion

√   **Organizational Problems**

- method (instruction)
- materials
- curriculum
- policies
- resources
- facilities
- staff
- customers

Once the categories are established, the procedure then is to brainstorm the possible causes, conditions, or factors under the appropriate category. Each of the ideas is placed on a horizontal line connected to the right and left of the category bone.

If it is an existing problem, in addition to brainstorming, there may be other data available. For example, if the problem is a copy machine constantly breaking down, there would be factual data concerning how it breaks down, how often, when it occurs, etc.

Like the brainstorming method, the Cause-Effect tool is easy to learn and can be done rather quickly. The key is to establish appropriate categories so that there can be focused brainstorming and data collection under each category.

The Cause-Effect Diagram (Fishbone Diagram) was developed to represent the relationship between some "effect" and all the possible "causes" influencing it. The effect or problem is stated on the right side of the chart, and the major influences or "causes" are listed to the left. The Cause-Effect Diagram on page 105 was developed to determine the root causes of disruptive students in the classroom.

The following seven-step process is an effective way of creating a Cause-Effect Diagram.

Cause-Effect Procedures

1. Draw a fishbone on a large chalkboard. A flip chart is satisfactory, but space will be restricted. Another option is to place two or more sheets of flip chart paper side by side on a wall or chalkboard. There is an advantage to this approach because the sheets can then be taken down and copied on a smaller scale.

2. Identify the problem to be solved and place it in the box at the end of the main bone arrow.

3. Agree on the categories to be used (four to eight) and make them appropriate to the problem being solved. The example on the next page shows how flexible and creative the categories can be.

4. Follow the brainstorming model to list the causes or conditions under the categories. There are two different ways this can be done. The first is to brainstorm each category in turn. The second is to use a more open process by placing any ideas suggested under the appropriate category.

   There will be times when an idea might seem to fit under more than one category; in this case, make a quick decision because the idea is more important than the category it is placed under.

5. Continue with the brainstorming model by clarifying and analyzing the ideas.

6. Determine whether there is one particular category causing most of the difficulties; if this is the case, prioritize the ideas according to which ones have the most negative impact and which can be changed or corrected. It is important to remember here the 80/20 rule. Eighty percent of the problem is usually caused by only 20 percent of the items.

7. If it is an existing problem, the solution arrived at can be verified by implementing the corrective action. If the activity was anticipating a potential problem, the verification can only be made upon implementation of the intended change or plan.

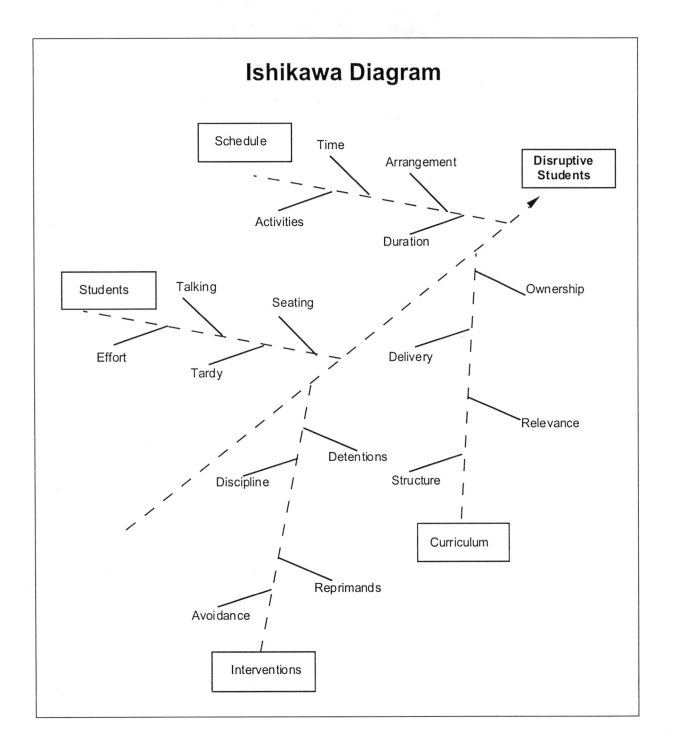

Cause-Effect Exercise

Review the material previously discussed and complete the following activities.

1. Review

Has the Cause-Effect tool been used in your organization?

_____

If so, was it found to be useful? (Please explain in detail.)

_____

_____

_____

_____

If not, why not? (Please explain in detail.)

_____

_____

_____

_____

Was it used appropriately and correctly? (Please explain in detail.)

_____

_____

_____

2. Practice

Identify a problem or issue and follow the seven-step process to determine likely causes for the problem.

PROBLEM STATEMENT:

_____

3. Evaluate

Was the Cause-Effect Diagram useful in uncovering all the potential reasons?

_____

_____

_____

_____

Checksheets

A Checksheet is a simple, efficient, and standardized method to collect data in an organized way for review and analysis. Sometimes the data being collected is for sampling purposes only, and at other times, it is an ongoing activity.

There are several important questions to ask and answer before deciding whether a Checksheet is needed:

• Why is the data needed?

• Is the data already being collected in some other form?

• How will it be used?

• Who will use it?

- • How will it be collected?

- • How often will it be collected?

- • Who will collect it?

- • What resources will be required to collect it?

- • What will it cost to collect?

- • Will the intended benefits exceed the value of cost and time?

One of the most useful purposes of a Checksheet is to detect patterns and to determine whether any problem, in fact, exists (a Checksheet does not solve a problem). The Checksheet will normally be used to construct a Pareto Chart, which will help to graphically portray the magnitude of the incidents being recorded.

The Checksheet on the next page is an example of how the tool is used to collect data.

Checksheet Procedure

1. Decide what data is to be recorded by asking the following questions:

   - • What is to be observed or monitored?

   - • Who or what does it, receives it, or is responsible?

   - • Where or how does it take place?

   - • When does it take place (time, day, month, how often)?

   - • How does it happen (how much, how long)?

   Note: "Why" is not asked because the purpose of the Checksheet is not to determine a cause or reason.

2. Design the Checksheet grid form to record the data to be collected.

3. Determine the time period or periods for recording.

Meeting Checksheet

ITEM	YES	NO
Room secured for meeting date?		
Room properly equipped?		
Meeting notices sent out?		
Meeting agenda developed?		
Meeting facilitator required?		

4. Ensure that there is honesty in collecting the data.

5. Assign the responsibility for recording, distributing, reviewing, and analyzing the data.

6. Develop a procedure to verify that the correct process was followed for collecting the data.

7. Determine what factors will be used to decide when a pattern becomes a problem to be solved or is an issue to be addressed.

8. Construct a Pareto Chart.

9. Target any problem it may identify.

10. If there is a problem, go through the problem-solving process; if there is no identified problem, determine whether the data collection should continue.

Checksheet Exercise

Review the material previously discussed and complete the following activities.

1. Review

Are any Checksheets currently being used? (Please explain in detail.)

_____

_____

_____

If so, is the information being collected effectively? (Please explain in detail.)

_____

_____

_____

Is the information being used appropriately? (Please explain in detail.)

_____

_____

_____

2. Practice

Identify some measurable data you would like to have or should have.

For example, what is the daily attendance rate of staff by week, by department?

Follow the ten-step procedure and answer all the questions.

Note: You may want this to be an extended exercise and actually collect the data for a period of time.

3. Evaluate

Determine how helpful the data is for the organization.

_____

_____

_____

Cost-Benefit Analysis

One of the factors to be considered in solving a particular problem or addressing an issue is Cost-Benefit Analysis. A Cost-Benefit Analysis, when properly done, provides an estimate of the real cost and benefits for a solution under consideration.

One of the characteristics of a quality organization is to estimate the total actual cost and the total benefits in relationship to the effort and energy required to institute a change. An example of estimating the total cost and benefit is the typical bidding process for a copier, wherein the lowest bidder is usually chosen with little, if any, consideration for costs and benefits over the estimated useful life of the copier.

However, there are some areas where it is more difficult to conduct a Cost-Benefit Analysis. The cost of implementing a new curriculum program with new texts and resource materials is easy to estimate, but it is more difficult to estimate the actual benefits against cost, compared to the current curricular program. Nevertheless, such a comparison must be made in order to justify the cost of the change and to receive support for it.

The finance department can be very helpful in doing some of the costing, and the staff involved in the curricular change could certainly list the benefits. The question that needs to be addressed as clearly as possible is whether the cost, in light of other priorities, provides for significantly greater benefits than those that exist.

A quality organization will always conduct a Cost-Benefit Analysis for every change process.

Useful Models, Tools, and Techniques

➜ **Management Information Systems**

Cost-Benefit Analysis Model

The following three-step process is designed to help you develop a Cost-Benefit Analysis model for your organization:

1. Calculate the hard costs of any change process. Include whatever is appropriate from the following:
   ◊ staff (use proportional costs where necessary)
   - salaries
   - benefits
   - substitutes

- personal days

- professional days

- meetings

- training costs (up front and long term)

◊ materials (original and consumable)

◊ replacement of lost materials

◊ equipment (net cost)

◊ installation (even if done in-house)

◊ supplies

◊ downtime

◊ operators

◊ administration

◊ inflation

◊ reimbursement

2. Once the costs are calculated, the next step is to determine the benefits. Factors to consider include the following:

◊ increase in productivity

◊ increase in efficiency

◊ increase in learning

◊ increase in attendance

◊ reduction in absenteeism

◊ improved product/service quality

◊ increased demand for product/service

◊ less turnover

3. The last step is to calculate whether there are, in fact, significantly greater improvements in relationship to cost.

Cost-Benefit Analysis Exercise

Review the material previously discussed and complete the following activities.

1. Review

Find examples (if any) of formal Cost-Benefit Analysis activities.

_____

_____

_____

If none, use the meeting's model exercise where it was costed.

Was the Cost-Benefit Analysis done correctly and honestly (use the three-step process)?

_____

_____

Was the Cost-Benefit data revealing?

_____

_____

2. Practice

Identify a change, program, or activity that was implemented recently and calculate a Cost-Benefit Analysis.

_____

_____

If you can't identify anything quickly, do a Cost-Benefit Analysis for processing a $5 purchase order.

_____

_____

_____

3. Evaluate

Is Cost-Benefit Analysis an effective way to help analyze waste, budgetary priorities, etc.?

_____

_____

_____

Force-Field Analysis

In every organization where changes are made or need to be made, there are always two critical forces at work. There are the driving forces (positive), which are concerned with implementing a change or improvement (and, in some cases, sustaining an improvement effort). In contrast, there are the restraining forces (negative) which are concerned with resisting a change effort.

The purpose of a Force-Field Analysis (see page 117) is to graphically illustrate the forces in a balance sheet format—the positive forces on one side and the negative forces on the other side. Once the elements of the two forces are identified, the goal becomes one of reducing the restraining forces and/or increasing the positive forces.

However, the problem is that all the forces are not equal nor are they all critical to successful implementation of the intended change. What needs to be done, once the forces are identified, is to determine which ones are most critical to success.

The process by which the Force-Field Analysis is developed is to simply brainstorm a list of the positive and negative forces with the group involved with the change process.

Note: Although this tool is usually used with a group, it can be used by an individual. Often it is an individual who begins to think about some needed change. Developing a Force-Field Diagram will help the individual to determine how best to "sell" the idea and to strategize how best to overcome the negative forces. This is a tool that is quick to learn and to complete. However, planning the strategy may take much longer.

Force-Field Analysis Model

The following nine-step procedure is used to develop a Force-Field Analysis:

1. Identify the problem or issue.
2. Discuss why it will be useful to identify the resisting and supporting forces.
3. Draw a "T" on a flip chart page, and at the top of the "T," write the problem or issue.
4. Have the group (those individuals who should be involved in the process) brainstorm the helping and hindering forces.
5. As a helping force (positive) is identified, draw a horizontal line up against the left side of the vertical line of the "T" and write the helping force on it. The hindering forces (negative) are placed on the right side of the "T."

Education
Force-Field Analysis

Forces for Change	Forces Opposing Change

- Tremendous pressure to reduce educational costs

                                      • Traditional education process

- More demand for dwindling financial resources

- Equipment purchases are the first items cut from budgets

                                        • Lack of knowledge of market environment

- Education is now a competitive environment

- Business is looking for leadership from education to help them cope with today's business problems

                                        • Short-term process

- Focus is to improve administrative and student outcomes

- Education must assume more responsibility for success of educational programs

                                        • Complacency

- Education must demonstrate added value to School Boards, Businesses and Community

- Change in education is slow and confusing

                                        • Lack of education

- Business and society are focusing on the poor quality of education as the cause of today's competitive business problems

                                        • Perceived lack of commitment on part of business

- Global competition is a major issue that education must face

6. Once all the forces have been identified, review the list and identify those forces that are internal to the organization (IO) and those that are external to the organization (EO).

7. Identify those negative and positive forces that are critical to successful implementation of the change.

8. Use the Mind-Mapping technique to develop detailed factors contributing to the negative or positive force. This will help to clarify what is meant by a particular force. For example, if "fear" is listed as a negative force, what are the factors contributing to "fear"?

9. Plan a strategy for enhancing the positive forces and/or reducing the negative forces.

Force-Field Analysis Exercise

Review the material previously discussed and complete the following activities.

1. Review

Has Force-Field Analysis been used before in the system?

_____

_____

_____

If so, was it found to be useful?

_____

_____

_____

_____

If not, why not?

_____

_____

_____

_____

2. Practice

Identify a change that will be taking place.

If none can be identified quickly, identify the forces for and against implementing site-based management.

Use the nine-step procedure.

3. Evaluate

Was Force-Field Analysis helpful and revealing?

_____

_____

_____

_____

Based on the forces for and against, what are the chances that implementation can be successful or be accepted?

_____

_____

_____

_____

List Reduction Techniques

One of the difficulties involved in using some of the quality tools and techniques in group situations is that many ideas and suggestions are generated, and unless there is hard, factual data available, it is sometimes difficult to reduce lists down to a manageable size or to one priority. If consensus can be reached, there is no problem, but this is not always possible.

The brainstorming technique, for example, allows everyone to participate, and it makes everyone feel that what he/she has to say or how he/she thinks makes a difference. However, how do you discard the ideas of some in order to come up with a workable list or reduced list and still have everyone feel okay? What is more important, how can a dominant force in the group be kept in check? Then, there is the reality that sometimes it is difficult for individuals to choose among alternatives unless a technique is available to "force" a decision.

Although there are several techniques to help accomplish the goal of reducing lists, the following two techniques will reduce lists quickly and fairly:

- **Nominal Group Technique**
- **Multivoting**

List Reduction Exercise

Review the material previously discussed and complete the following activities.

1. Review

What method or procedures have been used in the system to reduce lists created by brainstorming and other means?

_____

_____

_____

Have the Nominal Group or Multivoting Techniques been used?

_____

What have been the advantages and disadvantages of list reduction techniques?

_____

_____

_____

Which seems to be the most effective?

_____

_____

_____

Why?

_____

_____

_____

2. Practice

Use a previous brainstormed list or create a brainstormed list for any topic.

Follow the seven-step procedure for the Nominal Group Technique as discussed in the next section.

Using the same list, follow the eleven-step procedure for Multivoting.

3. Evaluate

Did one technique seem more effective than the other?

_____

If so, why?

_____

_____

_____

Do both techniques accomplish the purpose of list reduction effectively?

_____

_____

_____

Nominal Group Technique Model

The seven-step procedure below is followed to implement the Nominal Group Technique (NGT):

1. Review the generated list to be sure that similar ideas have been combined and refined.

2. Try to get the list down to 26 or fewer items, with an ideal target of no more than 15.

3. Put a letter of the alphabet in front of each item (in full view for everyone to see).

4. Give the group members a piece of paper and instruct each member to list along the left side of the paper the letters of the alphabet that were used to identify the items.

5. Count the number of items.

6. Instruct each member (working alone) to put the highest number next to the letter corresponding to the item of highest priority. If, for example, there were 15 items, the number 15 would be placed next to the item considered most important by the group member. The item considered the second highest priority would be given a 14 and so on until all the items have been ranked.

7. The sheets should then be collected from each member and tallied on a flip chart. The item receiving the highest score would be the first priority, and the rest of the list would be prioritized according to the scores. A decision can then be made as to whether only #1 will be dealt with or whether the situation calls for considering more than one item.

Multivoting Technique Procedure

The following eleven-step procedure is used to implement the Multivoting Technique:

1. Review the generated list to be certain that similar ideas have been combined and refined.

2. Place the items on a flip chart.

3. Assign a letter of the alphabet to each item (if there are more than 26, then use numbers).

4. Give the group members a piece of paper and have each one list the letters or numbers that have been assigned along the left edge of their paper. Instruct each member to place a #1 next to each item considered to be a priority.

5. Collect the papers and tally the scores.

6. Select the ten highest scored items.

7. Repeat the voting process using only the ten remaining items.

8. Tally the scores.

9. Select the five highest scored items.

10. Repeat the voting process using only the five remaining items.

11. The list of five remaining items would be prioritized according to the scores.

Matrix

One of the simplest tools to use is a Matrix Diagram, which is a grid of intersecting horizontal and vertical lines used to make comparisons. It is very quick to construct.

Action Plan Responsibilities Matrix

Individuals Assigned to Tasks

Action Items				
Awareness				
Communication of tasks and responsibilities				
Develop plan				
Resolve conflict with current policy				
Current school board policy				
Educate students and staff on new policy				
Measure plan's effectiveness				
Make necessary modifications				
Measure results				
Communicate result to all groups				
Standardize procedures				
Continued monitoring				

Matrix Construction Process

The following six-step process is used to create Matrix tools that can be used for a variety of activities:

1. Determine the standards or criteria to be used. A set of standards or criteria for staff development is shown in the illustration on the preceding page.

2. Determine which groups within the staff will need to meet specific criteria.

3. Construct the Matrix using the vertical column to show the standards or criteria and the horizontal column (across the top) to indicate the groupings of staff.

4. Start by using the first standard; then move across the grid and indicate under each staff grouping whether or not it is required or needed. This data could be solicited from a survey of the staff or from a staff development team that has identified development needs for different staff groups.

5. Analyze the Matrix to determine which staff groups need specific development.

6. Develop a plan and calendar to indicate when, where, how, and who will provide the necessary training.

Matrix Exercise

Review the material previously discussed and complete the following activities.

1. Review

How have Matrix Diagrams been used in the system?

_____

_____

_____

Have they proved to be helpful and useful?

_____

If not, why not?

_____

_____

_____

2. Practice

Identify an activity and develop standards or criteria and the related categories. For example, identify a series of problems to be solved (on the vertical axis) and identify the options available to solve the problems (on the horizontal axis).

Use the six-step process to construct the Matrix.

Analyze the information.

_____

_____

_____

_____

_____

Develop a plan or strategy to deal with the data revealed by the Matrix.

_____

_____

_____

_____

_____

_____

_____

3. Evaluate

Is this an effective way to display information?

Is this an effective way to use information?

_____

_____

_____

_____

_____

_____

_____

Pareto Diagrams

A Pareto Chart, also known as the 80/20 rule, reveals which causes among the possible many are responsible for the greatest effect. A general rule is that only 20 percent of the causes produce 80 percent of the effect—hence, the 80/20 rule. For example, 20 percent of the employees usually account for 80 percent of the absences.

As a matter of interest, the name "Pareto" comes from Vilfredo Pareto, an engineer-sociologist scholar, who developed this method in the 1800s.

The Pareto Diagram combines two of the common forms of graphs—the bar graph and the line graph. An example of a Pareto Diagram is shown on the next page. The bars are arranged in descending order of height (frequency) from left to right. This means that the categories represented by the tall bars on the left are relatively more important than those on the right. However, at times, it is very important to determine what is not considered a priority but may, in fact, be absolutely critical.

Pareto Charts are used to display the relative importance of all the problems or conditions in order to:

- determine which problem to solve first
- choose the starting point for problem solving
- monitor success
- identify a probable cause

Note: It should be recognized that the tools and techniques are not to be used in isolation from one another. For example, when the Pareto Diagram identifies a problem, a Cause-Effect Diagram can then be used to determine likely causes. This, in turn, could lead to another Pareto Diagram.

As the illustration below shows, the horizontal axis indicates the categories and the vertical axis indicates the numbers.

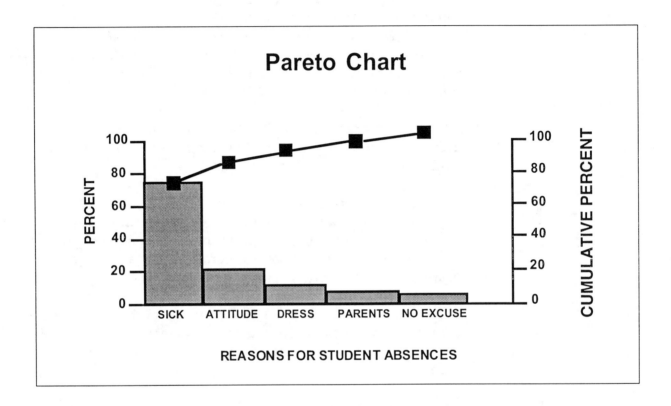

Pareto Analysis Procedure

The following seven-step procedure is used to create a Pareto Chart:

1. Determine the categories or classifications (days of the week, etc.) to be compared. The categories should range between 5 and 10, and an "other" or "miscellaneous" category should be used when appropriate.

2. Determine the unit or standard of measurement (frequency, cost, etc.).

3. Collect the data (if it is not already available). A Checksheet is a useful tool for this purpose.

4. Construct the graph by drawing a horizontal axis (dividing it into equal segments, one for each category) and a vertical axis at the left end of the horizontal axis. On the vertical axis, display the lowest to the highest count or frequency.

5. Plot the data by making a bar for the highest or largest category, starting at the extreme left side of the graph, and for each bar following in descending order.

6. The next step is to construct the cumulative (cum) line. This is done by starting at the top right corner of the first bar and extending the line to the right-hand corner of the second bar. The cum line is then extended diagonally to the top right corner of each succeeding bar; in essence, it will be in reverse order of the bar graphs. The purpose of the cum line is to ensure that the frequency of all the categories equals 100 percent.

7. Analyze the graph and use common sense. The most frequent or most costly events are not always the most important.

Pareto Exercise

Review the material previously discussed and complete the following activities.

1. Review
Has the Pareto Diagram been used in the system before?

_____

_____

_____

Was it helpful and useful?

_____

_____

_____

If not, why not?

_____

_____

_____

2. Practice

Identify available data that can be used to create a Pareto Diagram. For example, track student or teacher absences by day of the week and by school.

Follow the seven-step procedure.

Analyze the data.

3. Evaluate

Did the data displayed in this manner prove revealing?

What data should be tracked with the Pareto Diagram?

_____

_____

_____

Surveys and Interviews

Internal and external surveys (questionnaires) and interviews (face-to-face communication) are two important techniques in an educational environment because there is no clearly defined product specification.

The survey method is generally used when the number of people to be questioned is large, and the interview method is used when a small group is involved. However, when issues are of a sensitive or emotional nature, questionnaires will be more accurate even with a small group.

Although the other tools and techniques can collect a variety of measurable data, surveys and interviews provide valuable opinions, beliefs, concerns, and perceptions. In fact, perceptions are sometimes more powerful in influencing decisions than hard data.

Surveys and interviews are also useful in confirming information that may be the result of "rumors" or other verbal information. If, for example, the rumor is that morale is poor, a simple questionnaire, properly constructed, would help to determine whether it was true.

One problem with surveys and interviews is that the data is based on opinions and not necessarily facts, but they are an excellent way to start collecting factual data. Another problem is that both are time consuming in terms of preparation and analysis.

Surveys and Interviews Procedure

The following nine-step procedure is used to develop surveys and conduct interviews:

1. Decide what information needs to be collected, why it needs to be obtained, and from whom.

2. Determine whether a survey or interview is the best method to collect the information.

3. Select the appropriate method.

4. Construct the questions. This is by far the most important part of the process. Questions should be short, clear, specific, and easy to answer. Be careful to ask opinions on only one question at a time.

   Note: The questionnaire or interview questions should not be lengthy; two pages for a questionnaire and about fifteen minutes for an interview are reasonable guidelines.

5. Make the instructions clear if using a survey or follow a standard format for conducting the interview. It is helpful to pilot the questions first with a small group.

6. Design a procedure to conduct the survey or interview in an efficient manner.

7. If the interview method is being used, train the individuals who will be conducting the interviews.

8. After the survey or interviews are completed, collate the data.

9. Analyze the data.

Survey and Interview Exercise

Review the material previously discussed and complete the following activities.

1. Review

How have Surveys and Interviews been used in the system before?

_____

_____

_____

Were they effective?

_____

_____

_____

If not, why not?

_____

_____

_____

2. Practice

Identify some information that needs to be collected from a Survey or Interview.

Follow the nine-step procedure.

What are the results the Survey or Interview data?

_____

_____

_____

3. Evaluate

If an actual Survey or Interview was conducted, what was learned?

_____

_____

_____

How will the information be used?

_____

_____

_____

Could this same data have been acquired through some other means more expeditious than a Survey or Interview?

_____

_____

_____

Quality Planning

Proper planning is one of the two most crucial elements in the entire quality process of continuous improvement; the second is *executing* the plan.

Every organization has had "plans" that were developed with the best effort available, yet implementation was not effective. The prime cause for failure is that not enough quality time was put into the planning phase. A secondary cause is that implementation was not a priority, or, if it was, crisis management superseded the plan.

Therefore, every plan should follow a specific checklist procedure to ensure that no steps are missed and no shortcuts are taken, and it must be given priority.

Useful Models, Tools, and Techniques

→ **Brainstorming**

→ **Cost-Benefit Analysis**

→ **Force-Field Analysis**

→ **Matrix Diagrams**

Quality Planning Model

The Quality Planning Model is illustrated on the next page. It outlines a process for creating quality improvement initiatives in your organization. This process has been used in businesses, educational organizations, and government agencies in the United States and Europe.

Quality Planning Exercise

Review the Quality Planning Model and the material previously discussed and complete the following activities.

"Excellence in Operational Management"

Plain English	Stage and Focus	Total Quality Management
What are we about?	(a) Initially, Senior Management (b) Then as part of the improvement cycle, a continuous, recurrent, participative process	• Vision • Miaaion • Customer/Supplier Chain • Teams • Tools & Techniques
Who needs to do what to make this happen?	Progressively, all operational units of the organization from Board of Directors, Senior Managers to staff	• Functional Analysis • Process Modeling • Customer/Supplier Chain
How do we do these things?	As (2) above (i.e., everybody)	• Process Modeling • Process & Procedures • Customer/Supplier Chain • Cross Functional Teams • Conflict Management • Time Management
How do we know we're being effective in what we set out to do?	Board of Directors, Senior Management, staff, customers, and suppliers	• Setting Standards • Benchmarking • Assessment • Evaluation
How do we improve what we do?	Board of Directors, Senior Management, staff, customers, and suppliers	• Benchmarking • Redefinition of Standards • Process Analysis • Conflict-Time Management • Continuous Improvement Process

1. Review

Identify a plan that was recently implemented or brainstorm what usually goes wrong in the implementation of plans (changes).

Analyze whether it was implemented effectively. If not, what went wrong or was wrong?

2. Model

When another plan of any scope is developed, use the fifteen-step process in developing and implementing the plan.

3. Evaluate

Evaluate how successfully the plan was implemented.

Note: It may be appropriate to use the next exercise in conjunction with this exercise.

Process Decision Program Chart (PDCA)

One method of evaluating an implementation strategy is to use a simple procedure known as the Shewhart or Deming PDCA Cycle (there are variations of this procedure):

◊ PLAN (What to do?)

- goals and objectives
- description of the process
- resources required
- assigned responsibilities
- time lines
- benefits or expectations

◊ **DO (Doing it)**

- implement

◊ **CHECK (What happened?)**

- monitor

- measure

- evaluate

- analyze

◊ **ACT (What was learned?)**

- adjustments

- repeat the process

This cycle is usually used to pilot or field test a plan or procedure.

Galileo's PDCA Cycle

Based on our experience, business, education, and government professionals require a systematic approach to implementing quality. Therefore, we have modified and expanded the PDCA Cycle to provide our educational partners with a structured process for implementing TQM.

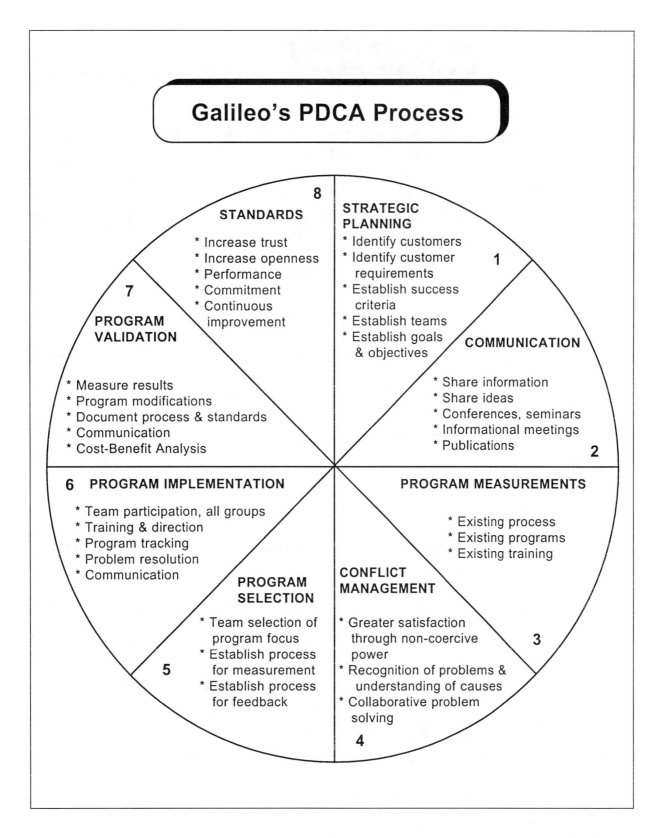

Galileo's PDCA Process

8 STANDARDS
* Increase trust
* Increase openness
* Performance
* Commitment
* Continuous improvement

STRATEGIC PLANNING 1
* Identify customers
* Identify customer requirements
* Establish success criteria
* Establish teams
* Establish goals & objectives

7 PROGRAM VALIDATION
* Measure results
* Program modifications
* Document process & standards
* Communication
* Cost-Benefit Analysis

COMMUNICATION
* Share information
* Share ideas
* Conferences, seminars
* Informational meetings
* Publications 2

6 PROGRAM IMPLEMENTATION
* Team participation, all groups
* Training & direction
* Program tracking
* Problem resolution
* Communication

PROGRAM MEASUREMENTS
* Existing process
* Existing programs
* Existing training

PROGRAM SELECTION
* Team selection of program focus
* Establish process for measurement
* Establish process for feedback
5

CONFLICT MANAGEMENT
* Greater satisfaction through non-coercive power
* Recognition of problems & understanding of causes
* Collaborative problem solving
3

4

PDCA Exercise

Review the material previously discussed and complete the following activities.

1. Review

What model(s), if any, have been used to track the cycle of implementing a plan?

_____

_____

_____

Has the PDCA Cycle been used in the system before?

_____

_____

_____

If more than one model has been used, which has proved most helpful?

_____

_____

_____

2. Practice

Identify a simple change to be made.

Follow the PDCA process. If no real implementation is to occur, go through the first two parts and then determine how part three (CHECK) would be done. It will not be practical to do the fourth part.

If no actual implementation is to take place, take advantage of the first opportunity where a change is to be made and use the PDCA Cycle.

3. Evaluate

In what ways does the PDCA differ from other models used previously?

_____

_____

_____

Which one was most effective?

_____

_____

_____

Quality Standards and Indicators

One of the most critical aspects in implementing a quality initiative is to develop quality standards and indicators (QSI). However, there is a strongly held belief in organizations with no tangible products that it is difficult, if not impossible, to measure what is done. This paradigm must change because the reality is "what gets done can be measured and what gets measured gets done."

The fact is that there can be no continuous improvement process without establishing QSIs as benchmarks against which success can be measured. The problem today is that business, education, and government professionals are being asked to be accountable, but there are no national or even professional "standards."

Therefore, it is up to each organization to develop its own QSIs. In doing so, it is vitally important that the stakeholders, internal and external to the system, be involved so that there is broad-based understanding and support.

Galileo has developed a Quality Indicators Master Checklist that can help to jump-start a continuous improvement effort. The purpose of this handbook is to provide the tools and techniques to identify problems, establish benchmarks, and measure progress. A sample Master Checklist can be found on the next page.

Every organization can measure what it does. Control or influence over the product or service may be limited in some organizations, such as schools and government agencies. What this means is that the job of these organizations is much more difficult and complex.

Quality Indicators Master Checklist
Survey of Selected Principles for a Quality Organization

Column #1—Personally agree with the principles. Scoring: 0 (strongly disagree) to 10 (strongly agree).

Column #2—Organization practices the principles. Scoring: 0 (doesn't ever do it) to 10 (always does it).

PRINCIPLES	#1	#2
1. Quality work and results are the priority	____	____
2. Delight external customers (employers, sending schools, parents, etc.)	____	____
3. Delight internal customers (school district staff, departments, programs)	____	____
4. Measurement is the key	____	____
5. Constancy of purpose	____	____
6. Fix the process vs. fix blame	____	____
7. Emphasize teamwork and cooperation	____	____
8. Build trust and respect	____	____
9. Data-based decision making	____	____
10. Focus on process	____	____
11. Constant education and training of staff	____	____
12. Continuous improvement (school district's practices and procedures)	____	____
13. Prevention vs. inspection (plan it right and do it right the first time)	____	____
14. Use of quality tools and techniques (to collect and analyze data)	____	____
15. Practice shared beliefs and values	____	____
16. Systemic planning and implementation (consider entire district when planning change)	____	____
17. All stakeholder groups involved (external and internal customers)	____	____
TOTALS	____	____

Quality Indicators Exercise

Review the material previously discussed and complete the following activities.

1. Review

Identify the quality standards and indicators currently being used in the organization in all operations and activities.

Start with your group by brainstorming a list.

2. Model

Develop a comprehensive list of quality indicators by using the eleven-step procedure (this will be a long-term process).

To simplify the exercise, develop a list for a department, process, etc.

3. Plan

Develop a plan to attain the quality standards.

Divide the plan into manageable and measurable parts.

4. Implement

Implement the plan.

5. Evaluate

Measure the results—one part at a time.

_____

_____

_____

Chapter 12

Seven Management and Planning Tools

The Japanese have gained a competitive advantage by developing a problem-solving process that combines the Seven Management and Planning Tools with the Basic Tools of Quality. The Japanese approach maximizes the strengths of these tools by using them to supplement each other.

The following is a problem-solving process that uses both the Seven Management Planning Tools and the Basic Tools of Quality. We call the process "Strategic Action Planning."

Step 1: Generate the Problem Statement

Without a rich problem statement, the team cannot be expected to do/complete its task. To carry it a step further, the team must know what the problem is before beginning to develop a solution.

Strategic Action Planning is dependent on the development of a rich problem statement. The problem statement is a tool to help the strategic planning team focus on what is really important. Personal commitment to resolving the problem is made by each team member.

Problem Statement Process

Team Problem Statement

- Each member takes two "Post-Its" and writes an idea for the vision statement on each one.

- "Post-Its" are placed on a piece of flip chart paper in a column down the left side.

- With all ideas considered, the team develops a vision statement defining success in resolving the problem.

- Each team member writes his/her personal commitment to resolving the problem.

- The team vision statement is written on the flip chart.

Output

- Your flip chart should look like the one on the next page.

Worksheet—Problem Statement

- Each team member receives two "Post-Its."

- Each team member writes two ideas that should be a part of the problem statement.

- Each "Post-It" is placed in a column along the left side of the flip chart.

- The entire team reviews all the ideas for consideration.

- The team, as a whole, develops a ten- to fifteen-word problem statement which will guide the team as it works toward agreement.

When the process is complete, write the problem statement at the top of the flip chart.

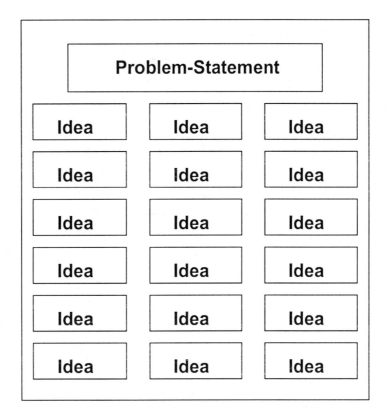

- Each team member writes his/her own personal commitment statement in support of resolving the problem.

Useful Models, Tools, and Techniques

→ Brainstorming/Imagineering

→ Affinity Diagrams

Affinity Diagram Process

- Each team member gets ten "Post-Its."

- In silence, each team member writes one idea per "Post-It" to answer the following question:

What are issues, tasks, or elements essential to resolving the problem statement?

Affinity Diagram Worksheet

Category A	Category B	Category C	Category D	Category E

- When finished, all "Post-Its" should be randomly placed upon the flip chart paper.
- In silence, team members gradually CATEGORIZE all the "Post-Its" into columns.

Step 2: Problem Area Selection

As the team's members group the ideas generated through the brainstorming and brainwriting exercises into categories, they are identifying the elements of the problem statement. This information is used in this step in the problem-solving cycle leading to the creation of a strategic plan.

It is reasonable to think that each category of actions would have an impact on the problem as a whole. It is also reasonable to conclude that all categories are not of equal impact and that there may be an interaction between categories. Completing actions or ideas in one category may resolve issues in another. The Interrelationship Diagraph will visually and logically draw out these relationships.

Useful Models, Tools, and Techniques

→ Reports

→ Checksheets

→ Interrelationship Diagraphs

Interrelationship Diagraph

Process

- On a clean sheet of flip chart paper, draw five boxes as shown in the example on the next page. Label each box with the title of the five categories from the Affinity Process.

- In each box write "IN" and "OUT."

- Look at categories A and B and decide if there is a relationship between them.

- If there is a relationship between these two categories, draw a line that connects them. If there is no relationship, there will be no line.

- By definition of this process, any relationship must be directional; that is, one category influences the relationship more strongly than the other, even if just slightly. Select the category that has the least influence. Draw an arrow at the end of the line that touches the least influential of the two categories. In the following example, there is a relationship between A and B, and B is a greater influence than A, as evidenced by the arrow pointing from B to A.

- Perform the same analysis between all of the categories you developed from the Affinity Process.

- Now count the number of arrows leaving each box and write the total in each box where you wrote "OUT."

- Do the same for the number of arrows coming "IN" to each box.

- The category with the most outgoing arrows exerts the most influence over the *system of ideas* you have developed.

- Select the three most influential categories to carry forward to implementation.

Output

You should now have a sheet of flip chart paper that looks similar to the example. Of course, your categories will have names that are a little more descriptive than A, B, C...

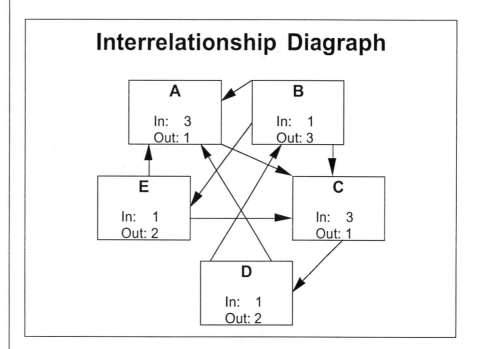

In the example, the three most influential are B, E, and D since they have 3, 2, and 2 outgoing arrows, respectively.

A strategic action plan that completely resolves the top three issues from the Interrelationship Diagraph will resolve about 80 percent of the *entire system of issues*.

The two issues not selected will be incorporated into the action plan for the most influential issues.

The final step in completing the analysis of the Interrelationship Diagraph is to identify your three most influential categories.

Step 3: Problem Specification

The Spider Diagram is a very useful tool for displaying the performance of multiple variables on a single page. The Spider Diagram makes it easy to read the data presented.

When a team is to determine what part of the problem statement is the priority that should be planned for and started first, it is often helpful to use both the Interrelationship Diagraph and the Spider Diagram together.

If the team were to use the Interrelationship Diagraph exclusively, it might choose the category identified as the "primary driver," concluding that by focusing effort in this area, other areas would be positively influenced. On the other hand, if the team were to use only the Spider Diagram, it might choose the category furthest from the vision established for it and thus work on the weakest area.

By using both the Interrelationship Diagraph and the Spider Diagram, teams can identify both the primary driver categories and those categories in need of immediate attention.

Spider Diagram

Process

- Determine the factors or variables that will be rated.

  Obtain factors from:

  Δ the same headers used on the Interrelationship Diagraph

  Δ performance factors from an organization

  Δ major vision elements for an organization

- Write the factors or categories on "Post-Its" and place them on the flip chart around an empty circle (see example on the next page).

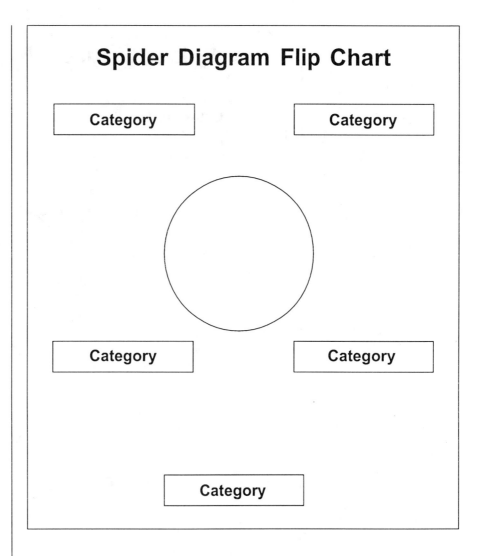

- Locate the center point of the circle and mark it with a pen. Then draw lines from the center point to each header on the outside of the circle.

- Using a scale from 1 (low, near the center) to 10 (high, near the outside), rate each category and place a mark on the associated line.

  Δ The center of the circle represents the worst possible performance, the furthest from the issue. The outside edge of the circle represents "perfection" or performance which is closest to the problem established for this category.

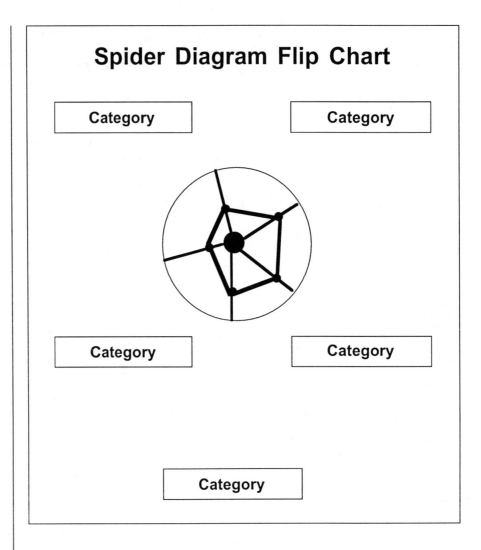

- Connect the dots on each category line to complete the "web."

- Evaluate the results of the display. The closer to the center the category rating, the more work must be done to achieve the vision for that category. Therefore, the category that is closest to the center should be addressed first.

Step 4: Determination of the Root Causes of the Problem

Tree Diagram

The Tree Diagram is a classic decomposition diagram. The problem is decomposed into its most basic elements using the data derived from the Interrelationship Diagraph.

Origins of the decomposition levels are:

- Level 1—Goal/Problem/Objective

- Level 2—The three most influential categories or key drivers affecting the goal/problem from the Interrelationship Diagraph

- Level 3—The ideas categorized in the Affinity Process

- Level 4—At least two steps required to implement the options in Level 3. You will develop this level using the Tree Diagram.

Process

- Write the Goal/Problem/Objective in the Tree Diagram, Level 1.

- Write the three most influential categories in Level 2.

- Write the best two implementation options for each category in Level 3.

- As a team, decide on at least two major steps required to implement each option and write them in Level 4.

- Steps must be quantifiable, measurable, complete actions that make a major contribution toward attaining the vision.

- An implementation step is something concrete. Thoughts or knowledge are not adequate. Each implementation step should result in tangible progress toward achieving the team vision.

Output

The output from this section is completed in the Tree Diagram Worksheet illustrated below.

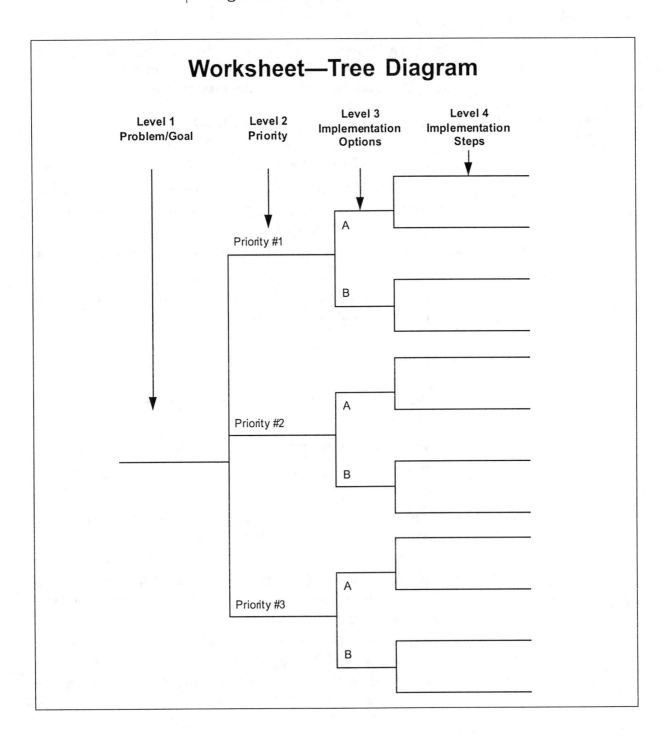

Step 5: Identifying Potential Solutions

Prioritization Matrix

The Prioritization Matrix is derived from the Tree Diagram. This is where orchestrating an Action Plan begins to take place. Sometimes we begin implementation based on what is easiest or maybe select an activity based on making a lot of progress quickly. Since you have built a system of activities, everything interrelates.

The approach of Action Planning is to take a holistic view of the vision, problem, priorities and implementation options, and steps. Without this perspective, it is difficult to sort out the issues to be addressed. Otherwise, you end up spending your time and resources doing the wrong things first. Without a holistic perspective, you may end up skirting the important issues to address less important issues just because they are more apparent.

You will be taking the implementation steps for all three of your priority items, mixing them together, and prioritizing them for the most efficient implementation.

Process

- Implementation steps from the Tree Diagram should be designated using letters.

- Each team member individually ranks each step in relation to all the others. If the letters are AL, the most important step is ranked 12, and the next is ranked 11. The least important step is ranked 1.

- Each member, in turn, calls out his/her rankings.

- Rankings for each implementation step are totaled in the proper column.

- Steps are now sorted, based on the totals, from highest to lowest.

Output

You will end up with a completed Prioritization Matrix, sorted from the highest to lowest priority step.

Prioritization Matrix Worksheet

In the Matrix below, each team member will rank the implementation steps from the Tree Diagram in order of importance. The most important will be given a "12" and so on down to the least important, which is given a "1."

Implementation Steps	Team Member	Team Member	Team Member	Team Member	Team Member	Team Member
A						
B						
C						
D						
E						
F						
G						
H						
I						
J						
K						
L						
M						
N						
O						
P						
Q						
R						
S						
T						
U						

Step 6: Developing Solution Strategies

Process Decision Program Chart

The Process Decision Program Chart is a method of evaluating and implementing proposed strategies. It is based on the assumption that something will go wrong as you move forward with your implementation steps. The PDCA Cycle enables us to develop a set of alternative actions.

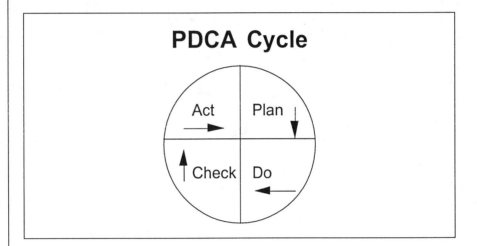

Process

Complete the PDPC tree-like diagram using the following steps:

- Enter the top three priority categories in Level 1.

- In Level 2, identify the three things that are most likely to go wrong or frustrate implementation of the action plan.

- At Level 3, identify three alternative actions that would put the plan back on track.

Output

Complete the PDPC worksheet on the following page.

Worksheet—Process Decision Program Chart

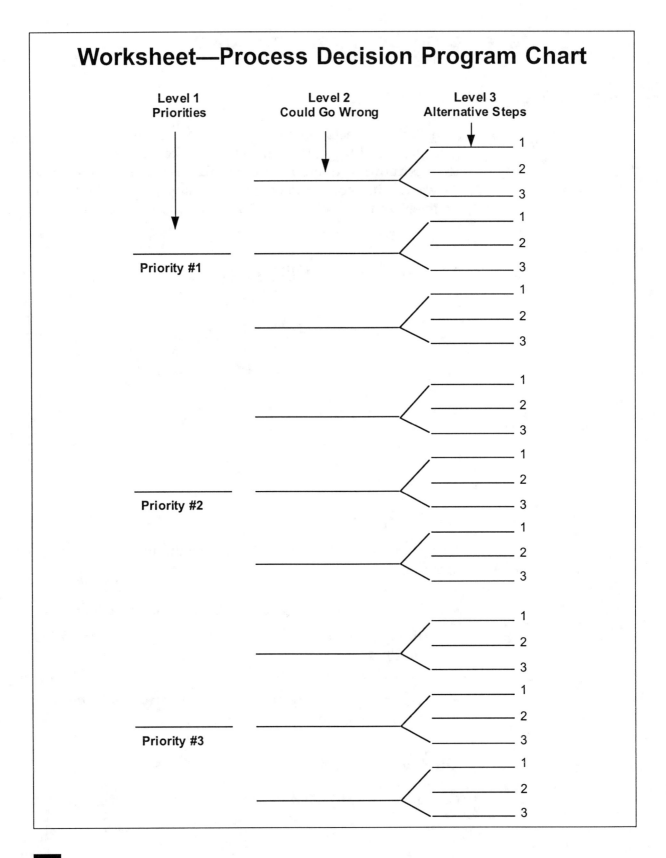

Step 7: Developing Implementation Strategies

Activity Network Diagram

The purpose of the Activity Network Diagram is to generate those final, specific plans for implementation based upon the desired, prioritized outcomes created by the project team.

Process

The process steps below are to be replicated for each of the prioritized outcomes that the team wishes to implement.

- Place two or three sheets of flip chart paper horizontally on a wall.

- Each team member should write down on "Post-Its" concrete, specific, measurable, and understandable **tasks** or **activities** necessary to complete the action plan that can be assigned to a responsible team member.

- Place the notes on flip chart paper.

- Arrange the notes in order of time sequence.

- On each note, write down who will execute that specific activity.

- The whole team gathers around to finalize the sequence paths.

- On each note, write down the estimated time it will take to complete.

- For each prioritized implementation option from the Affinity Process, the team should have an *Activity Network.*

> ## Output
>
> Complete the Activity Network Design worksheet below for each priority.

Worksheet—Activity Network Design

This is a process structured to guide development of the specific action steps necessary for the implementation of the three top priorities of an action plan.

Priority #1 _____ (top priority from the Prioritization Matrix)

ACTION STEP	WHO	HOW LONG	FINISH DATE

Process Evaluation

The team establishes a process for monitoring and evaluating the results. If the results are not achieved, the team returns to the first step and begins the cycle again. The team implements the following process:

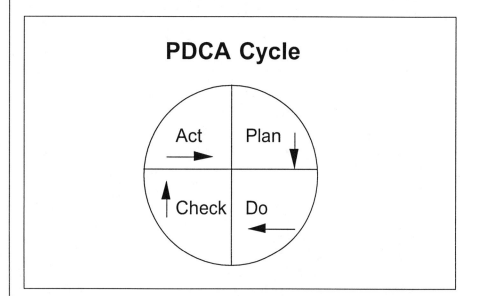

Chapter 13

Post-Course Self-Assessment

Name: _____ Position: _____

The following is a list of skills and knowledge covered in this workbook. If you feel reasonably comfortable with your level of skill or knowledge, place a check next to the item. If not, leave it blank. You should complete the self-assessment before using this workbook. It will help you to identify the topics requiring additional focus.

1. ___ Differences between facilitating, leading, and training

2. ___ Roles and responsibilities of the facilitator

3. ___ Roles people play in groups

4. ___ Adult learning principles

5. ___ Formation of groups

6. ___ Group member responsibilities

7. ___ Group problems and dynamics

8. ___ TQM principles and practices

9. ___ Decision making

10. ___ Union or association roles

11. ___ Conflict management

12. ___ Effective meetings

13. ___ Management presentations

14. ___ Problem-solving models

15. ___ Customer identification and relationships

16. ___ Brainstorming

17. ___ Cause-Effect Diagrams (Fishbone)

18. ___ Checksheets

19. ___ Cost-Benefit Analysis

20. ___ Force-Field Analysis

21. ___ List reduction techniques

22. ___ Matrix Diagrams

23. ___ Pareto Charts

24. ___ Surveys and Interviews

25. ___ PDCA

26. ___ PERT Charts

27. ___ GANTT Charts

28. ___ Listening

29. ___ Behavior observation and analysis

30. ___ TQM resources

31. ___ Team building

32. ___ Motivation

33. ___ Time management

34. ___ Flowcharts

35. ___ Histograms

36. ___ Mind Maps

37. ___ Run Charts

38. ___ Trust building

39. ___ Negotiation and mediation

40. ___ Strategic planning

41. ___ Organizational development

42. ___ Organizational diagnosis

43. ___ Process observing

44. ___ Affinity Diagrams

45. ___ Interrelationship Diagraphs

46. ___ Tree Diagrams

47. ___ Prioritization Matrices

48. ___ Process Decision Program Charts

49. ___ Activity Network Diagrams

Appendix

Worksheets

This section includes copies of the primary worksheets included in this course.

Facilitator Skills Assessment

Part I

According to the following scale, please rank your knowledge of and expertise in the following areas:

ITEM	LOW	MEDIUM	HIGH
Priority setting	___	___	___
Organization and planning	___	___	___
Multiple focus	___	___	___
Decisiveness	___	___	___
Communication skills—oral	___	___	___
Selection	___	___	___
Delegation	___	___	___
Organizational knowledge	___	___	___
Analytical thinking	___	___	___
Motivation	___	___	___
Fact finding	___	___	___
Goal orientation	___	___	___
Strategic thinking	___	___	___
Team building	___	___	___
Forecasting	___	___	___
Management efficiency	___	___	___
Systems perspective	___	___	___
Negotiation	___	___	___
Developing others	___	___	___
Risk taking	___	___	___
Leadership	___	___	___
Time management	___	___	___
Business management	___	___	___
Conflict resolution	___	___	___
Flexibility	___	___	___

Facilitator Skills Assessment

Part II

According to the following scale, please rank your knowledge of and expertise in the following areas:

ITEM	LOW	MEDIUM	HIGH
Accessibility	___	___	___
Analytical thinking	___	___	___
Assertiveness/autonomy	___	___	___
Attention to detail	___	___	___
Coaching	___	___	___
Communications—written	___	___	___
Conducting meetings	___	___	___
Control systems	___	___	___
Coordination	___	___	___
Decisiveness	___	___	___
Energy level	___	___	___
Evaluation	___	___	___
Feedback	___	___	___
Flexibility	___	___	___
Innovation	___	___	___
Meeting management	___	___	___
Motivating others	___	___	___
Networking	___	___	___
Organizing and planning	___	___	___
Participation	___	___	___
Persistence	___	___	___
Personality types	___	___	___
Quality principles	___	___	___

Facilitator Skills Assessment

Part III

According to the following scale, please rank your knowledge of and expertise in the following areas:

ITEM	LOW	MEDIUM	HIGH
Team preparation	____	____	____
Meeting management	____	____	____
Developing problem statements	____	____	____
Effective use of resources	____	____	____
Accuracy in forecasting	____	____	____
Timeliness	____	____	____
Follow-through	____	____	____
Training	____	____	____
Being a neutral observer	____	____	____
Defining the problem	____	____	____
Determining competencies	____	____	____
Creating a customer focus	____	____	____
Developing a team vision	____	____	____
Task focused	____	____	____
Critical thinking	____	____	____
Building expertise	____	____	____
Survival skills	____	____	____
Defining learning objectives	____	____	____
Developing training programs	____	____	____
Adult learning needs	____	____	____
Organization development styles	____	____	____
Keeping teams focused	____	____	____
Conflict avoidance	____	____	____
Valuing cultural differences	____	____	____
Assessment	____	____	____

Facilitator Classification Form

Please complete the following. It will help you identify your facilitator classification. You can also use this information to enhance your facilitation skills. You do not have to share this information with anyone.

ITEM	AREA OF STRENGTH	AREA FOR IMPROVEMENT
Facilitator's role and responsibilities	_____	_____
Different learning styles	_____	_____
Group leader roles and responsibilities	_____	_____
Group/team formation	_____	_____
Group/team roles and responsibilities	_____	_____
Group problems	_____	_____
Union or association roles	_____	_____
TQM/TQS basic principles and practices	_____	_____
Decision making	_____	_____
TQM resources	_____	_____
Organizing team meetings	_____	_____
Creating cross-functional teams	_____	_____
Developing team problem statements	_____	_____
Defining team issues	_____	_____
Creating a customer focus	_____	_____
Getting meetings started	_____	_____
Conducting effective team meetings	_____	_____
Resolving conflict	_____	_____
Developing management presentations	_____	_____
Adult learning theory	_____	_____
Motivation	_____	_____
Group dynamics	_____	_____
Sources of power	_____	_____
Team development	_____	_____
TQM/TQS models	_____	_____
Quality process models	_____	_____

Team Leader Roles and Responsibilities

Self-Assessment

ITEM	AREA OF STRENGTH	AREA FOR IMPROVEMENT
Act consistently	_____	_____
Eliminate fear	_____	_____
Develop trust	_____	_____
Give constructive feedback	_____	_____
Give praise and recognition	_____	_____
Provide whatever information is necessary	_____	_____
Model tolerance and flexibility	_____	_____
Create a respectful environment	_____	_____
Encourage creativity and risk taking	_____	_____
Keep the group focused on the task	_____	_____
Protect members from personal attack	_____	_____
Encourage participation and discussion	_____	_____
Be clear about how decisions will be made	_____	_____
Develop shared guidelines for group process	_____	_____
Encourage and support facilitator assistance	_____	_____
Follow guidelines for effective meetings	_____	_____
Evaluate group effectiveness	_____	_____

Team Member Roles and Responsibility

Self-Assessment

ITEM	NEVER PRACTICED	USUALLY PRACTICED	ALWAYS PRACTICED
Support and assist the group leader	_____	_____	_____
Participate by expressing opinions/feelings	_____	_____	_____
Support one another	_____	_____	_____
Maintain confidentiality	_____	_____	_____
Show loyalty to the organization	_____	_____	_____
Criticize constructively	_____	_____	_____
Suggest options or alternatives	_____	_____	_____
Adhere to the guidelines	_____	_____	_____
Do not allow personal attacks	_____	_____	_____
Do not monopolize the discussion	_____	_____	_____
Come prepared	_____	_____	_____
Keep the discussion focused on the task	_____	_____	_____
Provide data to support opinions	_____	_____	_____
Volunteer for extra assignments	_____	_____	_____
Listen to what other members are saying	_____	_____	_____
Try to understand diverse opinions	_____	_____	_____
Willing to "give and take"	_____	_____	_____
Ask questions	_____	_____	_____
Seek clarification	_____	_____	_____
Do not use inflammatory words	_____	_____	_____
Treat everyone with respect	_____	_____	_____
Do not bring hidden agendas	_____	_____	_____

Total Quality Principles

Self-Assessment

ITEM	NEVER PRACTICED	USUALLY PRACTICED	ALWAYS PRACTICED
Shared beliefs and values	_____	_____	_____
Shared vision and mission	_____	_____	_____
Teamwork and collaboration	_____	_____	_____
Constancy of purpose	_____	_____	_____
Consistent message and behavior	_____	_____	_____
Systemic improvements	_____	_____	_____
Prevention rather than inspection	_____	_____	_____
Data-based decisions	_____	_____	_____
On-going education and training	_____	_____	_____
Pride of workmanship	_____	_____	_____
Joy in learning	_____	_____	_____
Information sharing	_____	_____	_____
Problem solving	_____	_____	_____
Innovation	_____	_____	_____
Partnership development	_____	_____	_____
Customer focus	_____	_____	_____
Quality priority	_____	_____	_____
Fix the system	_____	_____	_____
Do not fix blame	_____	_____	_____
Process emphasis	_____	_____	_____
Cost-Benefit Analysis	_____	_____	_____
Value-added view	_____	_____	_____
Continuous improvement	_____	_____	_____
Elimination of waste	_____	_____	_____
Reduction of waste	_____	_____	_____
Valuing people	_____	_____	_____

Criteria Rating Form for Problem Selection

Problem Criteria	Problem Statement	Problem Statement	Problem Statement
Can the team solve the problem?			
Is the problem important?			
Does the team have control of it?			
Has the team agreed that it is a problem they want to work on?			
Total Score			

Ranking Key:
3 = To a great extent
2 = To some extent
1 = To a slight extent

Problem Selection Worksheet

Problem Statements ⟶	Problem Statement	Problem Statement	Problem Statement
Solve 1 2 3 4 5 Low High			
Importance 1 2 3 4 5 Low High			
Control 1 2 3 4 5 Low High			
Difficulty 1 2 3 4 5 Low High			
Time 1 2 3 4 5 Low High			
Resources 1 2 3 4 5 Low High			

Key: In the boxes across the top, write the problem statements the team is considering. Rate each problem statement against the listed criteria by working across each row. The higher the total score, the greater the likelihood that the problem is appropriate for the team to work on.

Solution Selection Worksheet

Solution Statements ⟶	Solution Statement	Solution Statement	Solution Statement
Solve 1 2 3 4 5 Low High			
Appropriateness 1 2 3 4 5 Low High			
Control 1 2 3 4 5 Low High			
Acceptability 1 2 3 4 5 Low High			
Time 1 2 3 4 5 Low High			
Resource Availability 1 2 3 4 5 Low High			

Key: In the boxes across the top, write the solution statements the team is considering. Rate each solution statement against the listed criteria by working across each row. The higher the total score, the greater the likelihood that the solution can be effectively implemented.

Facilitation Evaluation

Please rank the facilitator's contribution to the team meeting according to the following criteria:

	Excellent	Very Good	Satisfactory	Poor	Unsatisfactory
1. Did the facilitator/technical advisor have an accurate perception of what you needed and wanted?	☐	☐	☐	☐	☐
2. Were the objectives of the session clearly identified?	☐	☐	☐	☐	☐
3. Were the outcomes of the session clearly identified at the beginning of the session?	☐	☐	☐	☐	☐
4. Were the outcomes agreed to by the team?	☐	☐	☐	☐	☐
5. Did the outcomes match your expectations for the session?	☐	☐	☐	☐	☐
6. Was the sequencing of the session and material logical?	☐	☐	☐	☐	☐
7. Were you provided with the opportunity to discuss your ideas at the meeting?	☐	☐	☐	☐	☐
8. Did you get satisfactory answers to your questions?	☐	☐	☐	☐	☐
9. Did the facilitator/technical advisor use only language that is very clear to you?	☐	☐	☐	☐	☐
10. Do you feel that each step in the problem-solving process was satisfactorily explained to you before the next step was started?	☐	☐	☐	☐	☐
11. Was the facilitator/technical advisor easy to listen to and follow?	☐	☐	☐	☐	☐
12. Did the meeting achieve the stated outcomes?	☐	☐	☐	☐	☐
13. Are you satisfied with this session?	☐	☐	☐	☐	☐
14. Did the facilitator contribute to the success of the meeting?	☐	☐	☐	☐	☐
15. Would you recommend this facilitator for future use?	☐	☐	☐	☐	☐

Conflict Management Matrix

Departments, Organizations, Individuals

Areas of Conflict

Instructions: Rank the degree of conflict, on a scale from 1 to 5 with 5 being the highest, for each item listed in the Areas of Conflict column. Focus on reducing or eliminating those items with a high degree of conflict. Your program will be more successful if you can minimize the conflict of individuals, groups, or departments.

Areas of Conflict Matrix

Departments, Organizations, Individuals

Areas of Conflict				

Instructions: List the areas of conflict on the left side. List the stakeholders across the top (i.e., departments, organizations, or individuals). Rank the conflict according to how the stakeholder is impacted by the area of conflict according to the following scale: high, medium, low. Focus on resolving the areas of conflict for the stakeholder(s) with the highest degree of conflict.

Action Plan Responsibilities Matrix

Individuals Assigned to Tasks

Action Items				

Instructions: List the action items on the left side of the chart. List the individuals responsible for completing the action item across the top of the chart. Record the name of the individual(s) responsible for completing each action item in the appropriate box. (You can also include an optional completion date in the box.)

Quality Indicators Master Checklist

Survey of Selected Principles for a Quality Organization

Column #1—Personally agree with the principles. Scoring: 0 (strongly disagree) to 10 (strongly agree).

Column #2—Organization practices the principles. Scoring: 0 (doesn't ever do it) to 10 (always does it).

PRINCIPLES	#1	#2
1. Quality work and results are the priority	____	____
2. Delight external customers (employers, sending schools, parents, etc.)	____	____
3. Delight internal customers (school district staff, departments, programs)	____	____
4. Measurement is the key	____	____
5. Constancy of purpose	____	____
6. Fix the process vs. fix blame	____	____
7. Emphasize teamwork and cooperation	____	____
8. Build trust and respect	____	____
9. Data-based decision making	____	____
10. Focus on process	____	____
11. Constant education and training of staff	____	____
12. Continuous improvement (school district's practices and procedures)	____	____
13. Prevention vs. inspection (plan it right and do it right the first time)	____	____
14. Use of quality tools and techniques (to collect and analyze data)	____	____
15. Practice shared beliefs and values	____	____
16. Systemic planning and implementation (consider entire district when planning change)	____	____
17. All stakeholder groups involved (external and internal customers)	____	____
TOTALS	____	____

Affinity Diagram Worksheet

Category A	Category B	Category C	Category D	Category E

Interrelationship Diagraph Worksheet

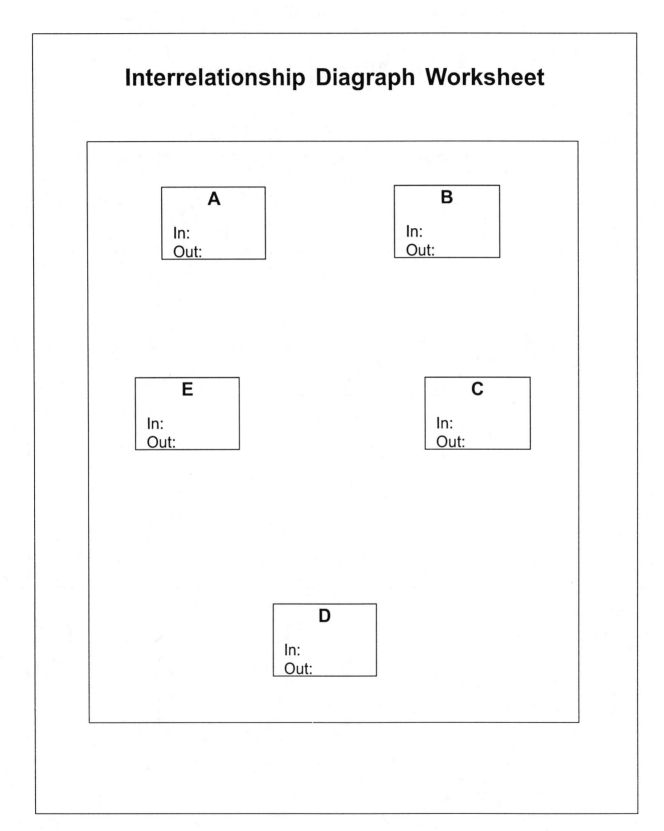

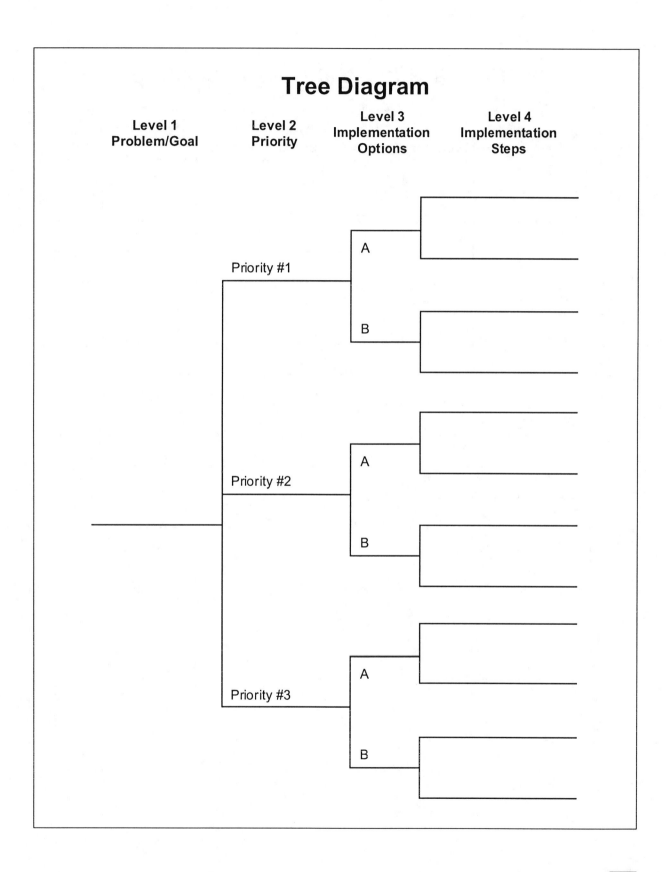

Tree Diagram

Level 1
Problem/Goal

Level 2
Priority

Level 3
Implementation
Options

Level 4
Implementation
Steps

Priority #1

A

B

Priority #2

A

B

Priority #3

A

B

Prioritization Matrix Worksheet

Implementation Steps	Team Member	Team Member	Team Member	Team Member	Team Member	Team Member
A						
B						
C						
D						
E						
F						
G						
H						
I						
J						
K						
L						
M						
N						
O						
P						
Q						
R						
S						
T						
U						

Process Decision Program Chart Worksheet

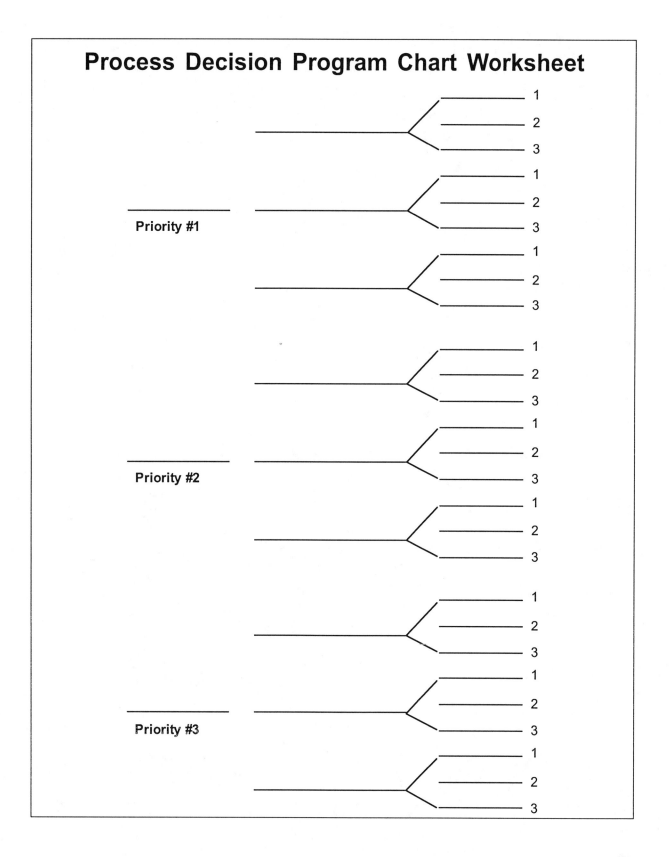

Priority #1

Priority #2

Priority #3

Index